Less Walk, More Talk

Less Walk, More Talk

How Celtel and the Mobile Phone Changed Africa

Russell Southwood

WILEY

A John Wiley and Sons, Ltd, Publication

Other Wiley Editorial Offices
John Wiley & Sons Inc., 111 River Street, Hoboken, NJ 07030, USA
Jossey-Bass, 989 Market Street, San Francisco, CA 94103-1741, USA
Wiley-VCH Verlag GmbH, Boschstr. 12, D-69469 Weinheim, Germany
John Wiley & Sons Australia Ltd, 42 McDougall Street, Milton, Queensland 4064, Australia
John Wiley & Sons (Asia) Pte Ltd, 2 Clementi Loop #02-01, Jin Xing Distripark, Singapore 129809
John Wiley & Sons Canada Ltd, 6045 Freemont Blvd, Mississauga, ONT, L5R 4J3, Canada

Wiley also publishes its books in a variety of electronic formats. Some content that appears in print
may not be available in electronic books.

British Library Cataloguing in Publication Data

A catalogue record for this book is available from the British Library

ISBN 978-0-470-74320-1 (HB)

Typeset in 11/15pt Times by Integra Software Services Pvt. Ltd.
Printed and bound in Great Britain by Printer Trento, Italy

Contents

Acknowledgements

At several points, those I have interviewed for this book have urged insistently that I do their story justice and "tell it like it was." Therefore I hope that this is not a bland corporate testimonial but a "warts and all" living story that reflects both the excitement and fears that were felt as it happened. For without mistakes, there is no learning and as the veterans of Celtel will surely attest, a life without mistakes is a dull life.

This is a book written from the recollections of both existing and former Celtel employees. Whilst it seeks to be accurate, we have not attempted to tell the stories of other mobile operators across the continent except where their paths have crossed with Celtel during the period described. Another person will write the more definitive volume that looks at the interplay between the major players once the dust has settled.

As this is a book that might be appealing to someone interested in either telecoms or Africa or both, it is also not a book that describes all that Celtel has done. It seeks to highlight only those things that were significant and important in its development.

I would like to thank the following who gave generously of their time when being interviewed: Tito Alai, Dr Saad Al-Barrak, Lord Cairns, Moez Daya, David Easum, Tsega Gebreyes, Rob Gelderloos, David Hagedorn, Gretchen Helmer (now Jonell), Mo Ibrahim, Omari Issa, Charlie Jacobs, Thomas Jonell, Mamadou Kolade, Kamiel Koot, Emilienne Macauley, Naushad Merali, Marten Pieters, Frederic Pichon, Lord Prior, Sir Alan Rudge and Tea Tuominen.

I would also like to say a big "thank you" to all those people who built up Celtel to what it is today, who are not mentioned above but who have or continue to play a role in its future development.

Special thanks must go to Rick Beveridge and Terry Rhodes for persuading me to write the book and to Martin de Koning for setting up all the interviews with both current and former Celtel employees and providing wise counsel during its writing. Lastly more than thanks must go to my wife Sara Selwood, who has supported me through the long process of writing this book.

The book refers to mobile operators rather than the American usage of cellular operators but the meaning is fairly clear from the context. More confusingly, the holding company went from MSI Cellular to Celtel at a certain point in the story. I have tried to refer to it as MSI up to the point where the name was changed to Celtel.

Introduction

A son wants to talk to his mother in the village. It is two days' walk each way.

Businessman Sani wants to set up a meeting with his colleague on the other side of Lagos. First, he sends his driver with a note suggesting a time. The driver hacks his way through the city's dense traffic before returning with another note suggesting an alternative time. In the event, it takes the driver two trips to confirm the meeting. The only thing Sani doesn't yet know is that his colleague will be delayed for an hour by a traffic pile-up on Ikoyi.

Abdillah is a plumber in Dakar. He spends all day making house calls on people before returning to the shop that takes his messages. He then rings all the people who have called wanting his services. Annoyingly, for him, over half of them have already gone out for the evening and as a result he misses talking to them. It can take him as many as three days to get back to somebody and in desperation customers sometimes drop by the shop that takes his messages. Those determined enough find out where he is currently working drive through the endless traffic jams of the city to see him. Abdillah is always short of money but never short of work. If only he could to talk to his customers...

Jackie likes to meet her friends for a drink after work in Nairobi. She phones round from work to see where everyone's going to gather. Everyone's a bit unclear about what they want to do except Sammy, whom Jackie finds a bit boring. So Jackie suggests a bar to meet at and hopes her other friends turn up. As it turns out, she spends an evening talking to Sammy, who is nice but dull. Her other friends decided to go to another bar but she only found out the next day. The man she will eventually marry is at the other bar.

A rural farmer wants to know what crop prices are in the market so he has to take a day out to visit one of the two markets where he might sell his goods. However, the prices are really too low at this point in the mid-week.

This was Africa before mobile phones arrived: a lot of walking and not much talking. In the main, fixed line phones were provided by a monopoly incumbent provider and they were not always either well resourced or energetic about connecting subscribers. What was particularly discouraging for subscribers at this point was that often the waiting list for new lines could be 10 years. But as always, it depended on who you were. In some countries civil servants got free phones and free calls. Shortages ruled and the speed of communication was more like a freeze-frame film than instant response. The absence of instant communications provided many excuses for things undone and little incentive for action.

In just over five short years, Africa went from having almost no phones to a position where over 100 million Africans now have access to a mobile phone. For them, life has become "less walk, more talk" as the phone makes their daily social and business routines much more efficient. With hindsight, this phenomenal growth was just waiting to happen. But the reality of early mobile roll-out in Africa was that the pioneers targeted the elites. It started with the "big men" paying hundreds of dollars a month to wander round with the latest status symbol, a large half-brick-sized phone. Everyone initially sought post-paid subscribers but found that getting them to pay after they had made their calls could be next to impossible. Mobile companies from the developed world simply thought that if this was the case they would stick to the few thousand reliable customers (who were a handsome cash cow) or retire from the field. It needed only one operator to offer pre-pay to light the touch-paper that set off the meteoric growth that was to follow.

At this point, Africa was considered a difficult place to do business and an even harder place to make money. As a result, almost all of Africa's first mobile operators were born from within the continent. As one of Africa's largest mobile operators, the story of Celtel's rise forms a significant part of this phenomenal growth. So this is not just an interesting corporate account but a first draft of a slice of the continent's now fast-moving history.

Celtel was founded in Europe by an African expatriate, the Sudanese Mo Ibrahim, who went from being the son of a humble clerk to running one of the continent's most successful mobile operators. This was a new development on a continent where current success in business could usually be traced to "connections" rather than merit alone.

For Celtel is not simply another multinational trading overseas. Its founder Mo Ibrahim came to Britain from Sudan and became a senior figure in UK mobile operator BT Cellnet before launching his own company. Against all the odds, he and the international team he put together convinced sceptical international bankers to invest in the company. His position as a man who operated "between cultures" allowed him to translate what was happening for those outside the continent. And far from being a basket case, Celtel demonstrated that the African continent was a place where serious money could be made.

Mo Ibrahim put together a truly multinational team with people drawn from all over the globe. At one point the senior management team included British, Dutch, Danish, Kenyan, Tanzanian, Ethiopian and Sudanese members, and across the company there were over 50 nationalities. A Celtel spokesperson in the Democratic Republic of Congo (DRC) in 2002 noted the diversity just within one part of its operation: "We find a Congolese as operations director in Celtel Gabon, a Zimbabwean as projects director in Kinshasa, a Mauritian as well as someone from Madagascar as regional directors in Mbuji Mayi and Lubumbashi."

Mo Ibrahim managed to inspire their enthusiasm and get them to work together to a common goal. More than that he made them believe that they could get difficult and seemingly impossible things done. One former employee confessed with regret that he wished he could get the same kind of commitment out of the staff of his new start-up.

The company also appointed African managers who were sent to run things in countries other than their own. It began a process of management development, sending these managers off to the London Business School to sharpen their skills. At a more humble level, the company was one of the major job generators over the last 10 years. It was a prime place to work and it paid well. As one manager noted: "When I arrived, everyone came to work on a bicycle or a motorbike. When I left, the car park was full of cars."

Those who work in mobile businesses often think of it as a business that is about getting the technical essentials right. The process of rolling out networks means that you need to have enough base stations to be able to offer the same or better coverage than your competitors. Marketing people will spend long hours devising promotions and giveaways to attract new customers and encourage more calling.

But in truth selling the ability to communicate in Africa, much as elsewhere, is also about the intangibles. Africa's pre-paid customers wanted the fancy phone, some even going as far as buying one whose full range of services could not be used. The mobile phone became the "sports car" of its age, something to impress those who you showed it off to. But unlike a sports car, this was an aspirational status symbol that almost anyone could use as long as they could afford the relatively low sum for their SIM card and first purchase of airtime. For those not able to afford a phone of their own, Africa's culture of sharing meant that for those on low incomes, as many as four people shared a single mobile phone.

Celtel went from being a start-up that could barely afford to bid for new licences to acquiring one of the continent's largest mobile operators in Nigeria for over a billion dollars five years later. It went from a handful of people with virtually no money, to a multibillion dollar company in just seven years! The untold story of Celtel is how it raised increasingly large sums of money to fund its growth in markets that believed Africa was a high-risk continent. In truth, Africa is as diverse as Europe with over 50 countries: these vary from places that have little or no corruption and almost no political risk to places where it is impossible to guess what might happen by lunch-time, let alone tomorrow. Against this backdrop Celtel was growing at a frantic pace and there was a constant race between the costs of running the operations and the potential opportunities for growth.

Although Africa is not entirely the land of civil war and famine seen nightly on TV screens across the world, it is probably one of the toughest places on the planet to do business. In order to acquire new licences, Celtel had to operate in countries that appeared marginal and gamble that the civil wars afflicting them would be resolved. Not every new business development executive gets to go to places where the hotels have no windows and there's no electricity or running water.

This book is the story of how Celtel grew its business with both successes and failures along the way: everything from launching a mobile network in the middle of a civil war to steering clear of corruption. This is no ordinary story about the corporate battle for market share in Europe and North America.

Growth in emerging markets like China and India may be impressive but it was much less surprising as it took off from a higher base of development in the respective economies. Africa is really the sleeping giant where mobile growth has been the electric force that has energised the emergence of a new kind of Africa. The fact that mobile companies like Celtel could put up and operate networks in countries where there were almost no roads showed that things could get done in Africa.

Celtel built networks and sold phones to consumers clamouring to buy their product. African consumers would in some cases literally batter the doors down to get their hands on a new mobile phone. On occasions the local police had to be called to control crowds of overenthusiastic customers. Sales targets in the thousands were recast in the tens or hundreds of thousands before being upgraded to the millions.

For all the many sophisticated ways there are of guessing when you put a business plan together, they all are in the end just "best guesses" until the market is mature enough to see most of the answer. Nigeria was a case in point. When the first "official" mobile licences were put out to tender, everyone knew it was a big market but a difficult country. Although Celtel lost out to others at this point, it was through this bidding process that most of the main players on the continent emerged.

In some ways the mobile phone business is like the oil business: politics and commerce are intimately intertwined. Licences are given by governments to operators and after the early years, the givers clearly understood the value of what was given. So in order to obtain licences or get things done to acquire licences, money would change hands. Celtel was remarkable in that it decided from an early point that it would steer clear of getting enmeshed in the devious ways of the "old Africa." How it negotiated its way through this minefield shows in practical ways that it is no longer enough to shrug and say: "Well, this is Africa."

But not all dealings with government were about money changing hands. Often Celtel was involved in high-profile deals that had wide-ranging political and legal dimensions. Celtel made legal history when it took the Tanzanian government to court in London over its threat to nationalise the company it had bought. There are few company executives that get to see whether a sovereign government will blink. It is also perhaps a tribute to the company's negotiating skills that the row was subsequently successfully resolved.

Celtel was a company that set out to "change the shape of the problem" in Africa. Instead of allowing phone calls between neighbouring countries to continue to travel 16000 km via Europe, it negotiated a deal that allowed a direct connection for a call between Kinshasa and Brazzaville, the two closest capitals in the world some 3 km apart across the Congo River. Previously the only direct connection was a ferry.

Taking the same principle, it launched a no-cost roaming scheme between three of its operations in East Africa. This was a global first and today extends to 12 countries, an area twice that of the European Union. Shortly after the first announcement, European Community officials rang the company wanting to know how it could be done.

In some cases it was too far ahead of its time. In Zambia it launched a payments company called Celpay that allowed its users to transfer money using mobile phones. It was also used to transfer money to demobbed fighters at the end of the civil war in DRC. Although the company began to develop a niche for itself, it was sold off when the company was preparing itself for its IPO. Now there are a number of new cash transfer launches from operators and this kind of service looks set to have a wider take-up.

It also developed Internet businesses alongside its mobile operations as it saw this as a natural development of its voice business. Again these were sold off in preparation for the IPO but subsequently, all the larger mobile operators have been scrambling to get into offering Internet services across their networks.

Celtel's coming of age moment was when it snatched the controlling stake in the Kenyan mobile operation Kencell after rivals MTN thought they had won. The winning deal was constructed on the basis of verbal agreements and trust, two qualities the continent does not always have in plentiful supply. The Kenyan deal was sufficiently large and the operation big enough to make other operators sit up and acknowledge that Celtel had really arrived.

Recognising the value that it had built, Celtel decided it would then raise further funds through an IPO. Inevitably other companies began to approach it about the possibility of a "trade" sale. After a last minute bidding war, its board agreed that the company would be sold to Kuwaiti-based MTC. This enabled Celtel to reach a new level in its development. MTC is now building itself as a global company and will rebrand all its operations as "Zain."

With the backing of MTC, Celtel was able to enlarge its position as one of the main African players by buying a controlling stake in Nigeria's V-Mobile. Mired in legal wrangling over shareholdings, Celtel was at last able to negotiate a way through claim and counterclaim and make an acceptable offer to the majority of its shareholders.

This book is written largely in a time sequence, with a beginning, middle and end. But much of what Celtel has done can only be understood by looking at key issues it has tackled. So, interspersed in the historical sequence of events are a series of thematic chapters that look at: how the mobile business works; dealing with corruption; the business innovations Celtel made; the success of its branding; and the hidden story of how it raised cash to fuel its growth.

The lessons from the rise of Celtel should be the case study material for anyone interested in business, globalisation and development. Emerging markets such as Africa offer the greatest growth potential to those skilled and brave enough to seize the opportunity.

CHAPTER 1

Nigeria:
bidding for the big one
in a land without mobiles

In January 2001 Abuja's Nicon Hilton was the setting for sub-Saharan Africa's first really big bidding process for mobile licences. A number of the continent's key players gathered around the hotel's pool as they waited for the auction to begin. Some of those coming to bid had surrounded themselves with bikini-clad young women so there was the slightly forced laughter of people trying to distract themselves. Underneath the casual light-heartedness, the observant might have noticed a certain wariness and caution. Nobody wanted their bidding strategy to leak to their rivals through an overheard careless word or two.

A similar bidding process in Morocco had attracted both considerable interest and money, so all eyes were on Nigeria. Everyone knew it was going to be a big market but few at the time realised it would become quite as massive. Nigeria had a population in excess of 120 million and only 1% was served by a phone of any sort.

But even if you were part of that 1% lucky to have service, it was far from reliable. One of those involved in the bidding complained that: "At this time, when you called (to an international destination) from Nigeria, 30% of your calls would not get through and 50% of the time the call would be hijacked by scammers trying to con money out of you."

MSI had already turned itself into a regional mobile operator of some standing but this was its first outing on its own in the big league. It was the point at which it came face to face with some of its later rivals, including a South African operator with which it was to have a close "love-hate" relationship over the years, MTN. After the bidding closed, it was clear from the final prices bid that whatever anyone might think about doing business in Nigeria, the African mobile market was going to be worth a whole lot more than analysts had believed. So although the Nigerian auction is not the beginning of this story, it is the first time that most of the major players walked on stage together.

The spur for the mobile licensing process was the widespread perception in Nigeria that the previous military regime had handed out licences widely as a way of getting income from bribes. As one local paper put it at the time: "...the past military administrations had caused the NCC (the regulator) to licence telecoms companies to provide GSM service...." The process was clearly lucrative as by the time the regime of the late General Sani Abacha had ended over 30 licences had been issued.

Among these were Celia Motophone Limited, Afritel Limited (a bidder in the auction), CIL, Reliance Telecoms Limited, United Networks Limited and Integrated Mobile Services Limited. Some of the companies were expected to offer national services, whilst others were going to provide regional services.

But of these known licences, only two were really operating: a consortium between a Lebanese construction company and local interests started Motophone and the Nigerian tycoon Mike Adenuga launched CIL with just over 20 base stations. The rest of the licences were in the hands of Nigerian wheeler-dealers who in the time-honoured "trading" tradition of the country were often said to be trying to offload these for more than they had bought them for.

The situation was completely chaotic and extremely confusing. One observer noted: "it was at a point where guys on the street would say to you: do you want a GSM licence? Everyone was supposedly validly licensed but it was unclear how this had occurred. Was it 'above board' or 'below the counter'? Nobody really knew what was going on."

In December 1999 the new civilian government of President Olusegun Obasanjo decided it would cancel all previous mobile licences and issue only four new licences. According to a government spokesman, this decision was based on the assumption that most of the licences were awarded to cronies of past military administrations and were not based on professionalism or the technical competence of the operators.

Also the government alleged that the process of awarding the licences had been, in most cases, fraudulent. Cynics might also argue that the government had seen the scale of fees generated by the licence auctions in Morocco and thought that if there's money going, we want our share of it. So bidders were given a deadline to come up with bids worth $100 million or more for these new licences.

At the time, those bidding for licences were of the view that the government and its regulator, the Nigerian Communications Commission, were finding it hard to slough off the old ways. A mixture of local and international companies turned up for the pre-qualification meeting. The locals included several of those, like CIL and Motophone, who were existing operators, and other hopefuls. Nitel and its subsidiary MTel bid separately as they could not agree on bidding together: as a result, MTel was unable to pay the $10000 pre-qualification fee. At an international level there were several companies of note: a group called Ideal-Levantis, a large Greek trading company in Nigeria that hooked up locally with its IT arm, Panafon, Vodafone's company in Greece, and MSI/Celtel.

Seven bidders were pre-qualified and an inter-ministerial committee appointed to evaluate the proposals. Few of the qualifying companies had ever operated a mobile phone network of any scale. Furthermore many of the pre-qualified were local companies but they did not have the necessary funds to pay the new licence fee. The unspoken assumption was probably that in the main they would be bought out by others, thus increasing the value of their investment for little extra effort. Neither the Ideal-Levantis consortium nor MSI/Celtel qualified.

David Easum, who was responsible for putting the bid together for MSI, was furious: "I caused a big stink and complained to the British High Commission, the US Embassy and (the West African regulators association) WATRA. I tried to get as much muck out as I could because the whole process was a joke. None of the qualifying entities could raise the required $100 million for the licence. This (lobbying) part went on for six months, then they decided that the whole process should be restarted."

The government learned from this fiasco, and the new bidding process was both well organised and transparent. The technical aspects were overseen by an international consultancy and there was an outside legal counsel present to ensure the rules were followed. In addition, it was the now independent regulator, the Nigerian Communications Commission, that was involved every step of the way rather than an inter-ministerial committee.

There were now going to be only four licences: three awarded to bidders in open auction and a fourth to Nitel's subsidiary MTel at the same price paid by the others in the auction. The reserve price was fixed at $100 million and a refundable deposit of $20 million was required, a condition that weeded out those unable to raise significant funds: local companies Intercellular and MTS failed to negotiate this hurdle successfully. The five companies that qualified to bid for the three licences were: MSI/Celtel, CIL, Econet Wireless, MTN and Orascom-backed United Networks. The auction of the three licences would continue till two of the five bidders dropped out.

On 16 January 2001 a mock auction was carried out at Nicon Hilton with bidders located in rooms along a corridor on the sixth floor in the full glare of the media spotlight. This rehearsal was to ensure that both those running the bid and those bidding clearly understood the rules of the game, as well as making clear to critics of the process that things were going to go very differently this time around. Furthermore the whole process was reported live on the Internet.

Those running the bidding announced they would be starting with a bid of $100 million but that they had the power to increase the bid price by as much as 50% after each round of bidding. At the beginning of each round a new higher price was fixed and the bidding companies had one of three options open to them: agree with the price, reject it or exercise one of their three available waivers. These waivers meant that the bidder could in effect wait out a round but once it had used all three of its waivers, it was out of the auction.

So the rehearsal day set the pattern for the auction. Government representatives arrived with a piece of paper saying that the opening bid was $100 million. The piece of paper had three options: yes, you agree to pay the bid price; no, you do not agree to pay the bid price; and lastly you can exercise your right to one of your three waivers. A short time later government representatives took away the bid paper with the response. It was subsequently bought back for signature by those authorised to bid and then everyone went away until the next round.

The five bidders represented most of the major players in the market. As there was not really much international interest in African mobile markets at this stage, all the players were from within the region.

When it bid in Nigeria, South Africa's MTN had only opened three operations outside of its South Africa base; the returns from those three operations, in Cameroon, Rwanda and Uganda, were modest.

However, unlike some other operators, it had spotted early on that African mobile markets were going to be primarily pre-pay, a factor that had enabled it to dominate the market in its operation in Uganda. It was hungry to break into new markets after a pause in its expansion, the potential in Nigeria was considerable and its business development team was geared up and ready.

Because of these factors or perhaps even despite them, it was significant that it secured one of the three licences available. It was a very profitable company and would have little difficulty raising the money for a high bid, although South African currency restrictions meant that it subsequently had to seek permission from the South African government to export such a large amount of currency to Nigeria to meet its payment.

Egypt-based Orascom was mainly a mobile operator in the Arab world, having secured licences in North Africa and the Middle East.

However, just before its IPO in 2000 it had tried to buy MSI/Celtel and, when that deal had failed to complete, it bought what it saw as the next most attractive acquisition target, Telecel.

Telecel in turn had been launched in the mid-1980s by Rwandan entrepreneur Miko Rwayitare and American pilot Joe Gatt. Together they had launched the first mobile network in Africa in the Congo. The two subsequently fell out and the company went up for sale, but as the management of the operation revolved very closely around its founders, Orascom was already beginning to discover that its acquired sub-Saharan operations were a mixed blessing and it later disposed of almost all of them.

However, when Orascom/Telecel came to Abuja it was hot on the trail of this major new licence opportunity and, following its listing a few months previously, had the money to keep up in the bidding process.

The other fancied winner of the auction process was MSI/Celtel, but in truth the company was going through a difficult phase. It had acquired numerous licences and was struggling to keep financing new growth. The company's founder Mo Ibrahim was at best ambivalent about doing business in Nigeria and in large part the board shared his scepticism.

Ten days before the auction MSI/Celtel had been using a mixture of market assessment and game theory exercises to decide what the other bidders might pay. From these it had worked out that the whole deal (including network investment) would cost $600 million and that it could afford to increase its bid to $250 million.

The perceived outsider was Econet Wireless founded by Strive Masiyiwa. It had started in Zimbabwe in the mid-1990s when Masiyiwa had very publicly taken on President Robert Mugabe and the government, and won, after a three-year battle that was taken to the highest courts in the country, the right to establish a privately-owned mobile operator. This legal challenge was only the first of several actions and he had already acquired a reputation as a man who was quick to take recourse in the courts. In the late 1990s, the Zimbabwean economy was sufficiently stable that he rapidly built up a successful business and used this as leverage to acquire a number of other African mobile licences.

Although it talked the game of pan-African expansion, the revenues from Zimbabwe and its other operations were proving to be insufficient to allow Econet to play in the big league unless it could find larger-scale financial backers. As a result, when it came to the bid, according to some of its local partners, Econet had difficulty meeting its share of the deposit and bidding expenses and these were paid by its Nigerian shareholders. Even at this stage, there were informal and confidential discussions between some of the local shareholders and MSI/Celtel about replacing Econet as their external partner.

The only local contender that made it through the more financially demanding second licence process was Mike Adenuga's CIL. It was thought that Adenuga's fortune had been made under the previous military regimes in the fields of oil and banking and he was determined to get into the new boom sector of mobile telecoms. His existing operation already had 30000 subscribers in Lagos and he said that it would be possible to increase this to 150000 subscribers very quickly. To add external expertise, he went into partnership with Deutsche Telekom's international consultancy Detecon. With the ebullient confidence for which Nigerians are so famous on the continent, he was reported to have said that the only way the process would be over and for him not to have a licence, would be for him to be dead.

The most obvious absent bidder at this point was South Africa's Vodacom. It had sealed an agreement with its minority shareholder Vodafone that it would not compete north of the equator and it was only later that it decided to push for acquiring a more continental presence. Also France Telecom (later to rebrand itself as Orange) was not yet interested in Anglophone countries.

On the morning of the bidding the five people allowed from each of the teams came out of the lifts and stood outside the doors that led to the corridor where the auction would take place.

The Orascom and MSI/Celtel teams nodded at each other and the MSI/Celtel team commiserated with Naguib Sawiris, the chairman of Orascom – Orascom had just launched their "Oasis" network in the DRC the week before whilst MSI/Celtel had had theirs up for about six weeks and both were concerned about their people in Kinshasa: news had just come through of the assassination of Laurent Kabila, President of the DRC, the day before.

At the agreed time the people running the auction ushered the teams along the corridor and into their allotted rooms where they would be closeted without phones for eight hours a day whilst the bidding took place.

The bidding process only allowed six rounds a day and so the 17 rounds that the actual process took lasted two and a half days but did not go entirely as MSI/Celtel's gaming exercises had predicted. It had no idea what

Mike Adenuga would pay as his determination to get a licence might easily lead him to overbid. It guessed that Econet would not get funding for more than a US$200 million bid and that Orascom would not go over US$180 million. On this basis, it expected the bidding to top out at somewhere not too far north of US$200 million.

For security reasons, MSI/Celtel's bid limit of US$250 million was only known by three staff members, two on the team and one at head office. It was conveyed in code to the team members because the phone lines were considered to be insecure. And the local Nigerian investors were also kept in the dark. Bid team member Rick Beveridge said: "We didn't tell the local partners, or even fellow team member David (Easum), until we were halfway through the process."

The bidding started slowly going up in tranches of US$10 million but quickly picked up momentum as the auctioneers realised there was a considerable financial appetite for the licences on offer. Rick Beveridge remembers: "it was how they describe flying an airliner – long periods of boredom punctuated by periods of intense activity. We would work for five minutes and then chat for a couple of hours – I remember that we wondered whether we could persuade the auctioneers that the MTN girls by the swimming pool six floors below were communicating with their team members by some sort of code."

Bidding didn't pass the US$200 million mark until late on the second day of the auction and MSI/Celtel's team were bullish about the outcome. The Orascom-backed team had put in their first waiver at $180 million and there was a strong feeling that things would close out very quickly on the third morning.

The MSI/Celtel bid team's confidence was not shared by senior management in Europe who were contemplating the Olympian task of raising $600 million, when every sinew of the company was straining just to remain upright. Even bid team leader Terry Rhodes was unsure: "it's debatable that if we'd we won that we could have found the money. We had 14 days to do it if we had. I kept saying to everyone, 'if we win this, you have to come up with the money in 14 days.' I knew they weren't taking me seriously."

The MSI/Celtel team and the local investors took their satellite phones (that were considered secure) up to the roof of the hotel and under a star-speckled sky conducted a series of tense conversations with MSI's founder and its CFO. Founder Mo Ibrahim could see the bigger picture and was worried about the consequences of winning. He was phoning the CFO and saying: "Why haven't they dropped out?" Terry Rhodes believed that if they lost the bid at a high price it would keep MTN occupied in management terms for two years, giving MSI more of a free run at the rest of Africa. But at the time each side was pressing hard to get its own way.

Mo Ibrahim felt he stood between his "rottweilers" and the board: "There was an intense debate at board level. We had a cash crunch at the time and if we'd won we would have had a big problem. We always tried to bite off more than we could chew but my job was to make sure we didn't suffocate on it. I had to keep my managers who were like my rottweilers hungry but I needed not to destroy the board. There were a lot of white hairs around the table during this process."

On the third day, the auctioneers started the day by proposing the maximum possible under the rules, $248 million, right on MSI/Celtel's bidding limit, but bidding then rose to $265 million, well above MSI/Celtel's predicted outcome. MSI/Celtel started using its waivers, staying in the bidding but not committing itself to pay that price. Orascom did not use any of its two remaining waivers, but went straight out – its entire team left the hotel, drove to the airport and flew out of Abuja immediately. Caught by local journalists at the airport, its banker responded angrily that the whole process had been a joke, and the licences were not worth this sort of money.

As the bidding moved above $250 million, MSI/Celtel's strategy was to push up the price by implying it might come back in. It used all three of its waivers and as the numbers rose, both Econet and MTN exercised their first waivers. Determined to win a licence, CIL said "yes" again and again as it had in every previous round. It was almost as if it was not only determined to win but to show that it could pay the highest price.

Watching on the Internet from Europe, with no way to contact the team locked in the room in Abuja, Mo Ibrahim was calling CFO Kamiel Koot with ever-increasing agitation, asking why the team had not followed Orascom and dropped out entirely.

Eventually, at $300 million, MSI/Celtel had used up all its waivers and had to say "no" and dropped out. Under the rules the price to be paid was the highest price where the remaining bidders had all replied "yes," which was a whopping $285 million, a great deal more than probably any of the bidders thought they would end up paying. The winning bidders were CIL, Econet and MTN and the "consolation prize" licence went to Nitel's MTel, which also had to pay the same high licence fee. Although there was much celebration by the winners after the auction, only one of the victors – MTN – reaped the initial rewards they expected.

Robert Nisbet, MTN's finance director of the group, acknowledged that the "price is a bit on the upper range, the company is happy to have won the licence and will concentrate on building a proper and reliable network." He said his company was looking forward to the next 10 years during which it intended to invest $1.4 billion.

The first of the other bidders to go down was CIL. Having paid a $20 million deposit, bidders were expected to come up with the balance of the licence fee just under a month later on 9 February 2001 or forfeit both their licence and their deposit. The money had to be transferred to the NCC's bank account in New York.

At close of business on the day, the NCC signalled in an announcement that it had not arrived: "…(the) balance of the licence fees due from Communications Investment Limited, CIL, had not cleared into the NCC account at Chase Manhattan Bank, New York, despite the earlier advice received from CIL that the payment of the relevant fees had been effected. Communications Investment Limited has therefore not fulfilled its payment obligation as stipulated in the Information Memorandum."

CIL's explanation was that it was willing to make the payment but on the condition that a clash over the operating spectrum was cleared up: "in the course of arranging for the payment in question, CIL's financiers were seriously concerned over the fact that the frequency allocated to CIL by NCC is currently the subject of a litigation which could lead to its suspension; with serious repercussions for the operational targets of the company." The company's spokesman, Harry Willie, added that in order to get this sorted out, CIL was compelled to place a few days' moratorium on the withdrawal of the funds in order to seek clarification from the NCC whilst a senior official was sent to the NCC headquarters in Abuja to discuss the matter.

Harry Willie stated that "the fact, however, is that CIL had indeed transferred the entire balance of the licence fee to the designated account by the deadline date of 9 February 2001 and a letter advising NCC of the transfer was faxed to the NCC on the same day along with a copy of the bank transfer statement." However, the letter made clear that NCC's ability to collect the payment was conditional.

Because it had committed itself to a fair and transparent process in which everyone had to abide by the same rules, the NCC was in no mood to be bargained with. Cynics were already saying that CIL was playing for time. An NCC spokesman at the time described the reason given as an excuse since it was not raised when the frequency was initially allocated. When the NCC board met, it decided that CIL had not met the terms of the auction and should forfeit its deposit and be excluded from obtaining any new licences for a period of five years.

MSI/Celtel sought to pursue the unused licence, saying in effect "we're next in line" but was told that the licence was going to be shelved for the time being. One unintended consequence of local shareholders raising dollars for their share of payments was that the Nigerian currency, the Naira, fell from N124 to the US dollar to N131. Local currency markets had never had to raise this scale of foreign exchange in just a week.

Despite its five-year ban, just two years later CIL's owner Mike Adenuga came back as Globacom and took a broader operator licence that included a mobile operation. The cost of that licence? Just $200 million. This was the reserve price set by the NCC and as Globacom was the only bidder, it secured the licence at that price. Some guys have all the luck.

Econet's Strive Masiyiwa subsequently fell out with many of the local shareholders in Nigeria and control of the company fell victim to a complicated legal action, parts of which are described later. MTel's licence fee costs simply added to Nitel's burgeoning collection of debts without the company ever realising sufficient market share.

For MSI/Celtel, the Nigeria auction was the right deal at the wrong time but it was later to return in triumph. Board member Lord Prior's version perhaps best sums up the result: "The board authorised the bidding to go up to $250 million. We were a bit worried. We thought Nigeria was a good thing but a bit corrupt. If we won, we didn't know how we could have raised the money. It would have been a mistake not to go for it but if we'd won, we'd have gone from 'the frying pan into the fire.'"

As it was, MTN soon found itself as the only well-financed player in the Nigerian market and using all its resources just to try to keep up with the incredible pent-up demand. As Mo Ibrahim and Terry Rhodes had predicted, they effectively removed themselves from bidding for new operations for more than two years, giving MSI some space to hone its business model and build other operations elsewhere in Africa.

CHAPTER 2

The boy from Sudan makes good

Some lives are predictable; others are less ordinary. Celtel's founder Mo Ibrahim went from being the son of a Nubian clerk in Egypt to running one of the largest mobile companies in Africa. But this journey to fame and fortune could probably not have taken place in the country of his birth. It is an extraordinary journey that includes both foresight and luck in equal measure.

Mo Ibrahim's father was a clerk who came from a village in Nubian northern Sudan to work for a small Italian cotton export company in the Egyptian port of Alexandria. Mo Ibrahim was born in the Sudan in 1946 before the family moved to Egypt. Ibrahim remembers his father's role as was pretty lowly: "When the cotton was bought, he used to take the money to the bank. It was a blue collar job and it was not much."

It was perhaps his Nubian roots that marked Mo Ibrahim out as different and special. Nubians can be found in both Egypt and Sudan; they speak their own language and have their own culture and heritage. They have a reputation for honesty and speaking their own mind. As Mo Ibrahim said: "all this gives you an element of specialness. These are deep roots and it gives you stability and a sense of place."

This sense of specialness gave Mo Ibrahim a strong self-confidence which has served him well throughout his life. This self-confidence was combined with the kind of intelligence that propelled him to the top of his class and off into the next phase of his life as an electrical engineering student at the University of Alexandria.

He attained a BSc in Electrical Engineering before returning to his native Sudan to work for the government-owned telephone company which at that stage was called Sudan Telecom. However, before long he was off again in search of more education, this time to the UK to take an MSc in Electronics and Electrical Engineering at Bradford University for which he obtained a British Council scholarship.

Having secured his MSc, his next move was to take a PhD in Mobile Communications from the University of Birmingham. In 1974 mobile communications did not really feature mobile phones but the area Mo Ibrahim chose to examine was to become the heart of the new mobile business: what happens to a transmitter and a receiver when one or both are moving. In studying this, he looked at how things like buildings interfere with signals and drew up equations that sought to explain the underlying physical laws of mobile wireless signalling. His life could not have been more different from the tense, incident-packed pattern of his later life in business: "I was a researcher at the University of Birmingham. A scientist, a person who works for glory, who is capable of spending hours doing experiments and calculations!"

He was far-sighted in his choice of subject as there were no mobile phones at that point and even in academia, discussion of mobile phones was kept within certain limits. As Mo Ibrahim remembers: "My thesis was on the feasibility of mobile communications. I was almost failed. One of the external examiners was unhappy that I suggested using high frequencies (900 MHz)

Mo Ibrahim

as these were not available. It was seen as heresy." (These frequencies now support over two billion mobile phones worldwide.)

But his curiosity about mobile technologies had been aroused at a much earlier point. In 1969 he was working as a young intern at the International Telecommunications Union in Geneva and had seen a mobile wireless system used in taxis: "I saw some guy in a taxi using a private mobile radio system and I was fascinated by it. How does it work? How does it go through valleys and buildings? It was a line of sight communications technology."

A year before his next job at a British Telecom company called Cellnet, he was at a trade fair in the USA: "Motorola's engineers presented me with a prototype of the mobile phone. They lent it to me for a day, and I immediately understood the interest: the telephone was no longer linked to the house, the office or the car. It belonged to an individual, to me. Sure, the handset was rather heavy and cumbersome, a bit like those walkie-talkies that you see in certain American films on the Second World War, but it offered so much more!"

However, the Cellnet that Mo Ibrahim joined as technical director had little or nothing to do with mobile phone networks as we know them today. The company was started in 1983 to pioneer an in-car telephony service and was supposed to start operations two years later. At that time, mobile phones used analogue rather than digital transmission. The handsets were heavy and awkward to use, needed large aerials and used lots of energy. Back then a phone needed 20 watts to power it compared to just a few watts used by today's phones.

Mo Ibrahim led the design and deployment of the first mobile phone networks in the UK in 1985. Starting a mobile network was not like anything BT had done up to this point: "We had to do a complete review of the network that we were in the process of installing. It needed aerials not only for roads, as planned for car telephones, but also on roofs, in order to allow telephone calls to be made from the building opposite, for example. We also needed to make sure that the handsets were available. We ordered 5000 handsets. This was the first mass production of such a device … And this was how, in London, in 1985, the first mobile network in the world began operating."

Terry Rhodes remembered his regular meetings with Mo Ibrahim from the time when they both worked in BT Cellnet: "I was doing business planning and capital expenditure. I often had to talk to the technical people and there was Dr Ibrahim. He would always arrive late and smoke a pipe. We had long talks about where the business was going in his grubby office in the Elephant and Castle."

As soon as BT started its networks, other networks started operating and both equipment manufacturers and operators came together in a process that led to the creation of the GSM standard that is now known throughout most of the world. In 1987 the European Commission succeeded in getting all the states to agree on the same standard for a mobile telephone that was to be both digital and cellular. Each member country committed to using it. In this way, the new products accessed a market of several hundred million customers in (the then) 12 countries, instead of being limited to a single country and 50 million inhabitants, for example. And it was from this European base market that GSM went out across the world.

Having just turned 40, Mo Ibrahim decided that it was time for a new challenge and went off to set up his own company. By the early 1990s, Margaret Thatcher's Conservative government had decided to privatise a number of previously government-owned companies and to open up the economy – particularly the telecoms sector – to a greater level of competition. By this stage, Mo Ibrahim was fed up with working for a big company "which was too complex and too frustrating for my taste. I wanted to be my own boss and decide my own fate."

The new company was called Mobile Systems International (known as MSI) and the business opening that Mo Ibrahim had spotted was that in this new era of competition, there were various new players in mobile telecoms around the world and many of them had never had anything to do with telephony before. He offered advice to these new players on installing their networks and secured his first contract in Sweden. A year after starting the company he had hired 10 engineers and by 1992 there were 25 employees.

MSI developed the software which captured and automated the previously "dark arts" of designing a radio network. This software package, called Planet, was sold from 14 offices worldwide and used by many major operators and network suppliers. Amazingly, it is still in use today; Mo Ibrahim reckons it has been used to plan about one-third of all GSM networks in the world.

By the time MSI was sold to Marconi in 2000, it had 800 employees. This was Mo Ibrahim's first start-up and in many ways was to be both the experience he learnt lessons from and the template for his later venture, Celtel. Mo Ibrahim proved to be an unusual entrepreneur who did not fit the usual stereotype. Where other start-up entrepreneurs might have a tendency to micro-manage, he was perfectly prepared to delegate. Also where others might be greedy, he was prepared to share the profits through granting shares to his employees.

These employee shares were not unique during this period but were a key part of MSI's ethos. He either gave employees shares when they were hired or later as a form of bonus. So over the life of the company before it was sold in 2000, the staff owned 30% of the company. Shares and share options were a relatively new development and employees were often fairly sceptical about them. Mo Ibrahim would explain them as a form of additional reward designed to thank staff for their time and work. The value of the shares was usually scaled against the size of salary paid. Initially a value of 14p was given to each of the shares through a process of comparison with other companies. When the company was bought by Marconi, employees were able to sell their shares for £14 each, 100 times more than their original value.

The same process of reward was repeated for Celtel employees: "For employees, it was a bit like an annual bonus. I repeated this form of operating within Celtel, and, when we were acquired by MTC, the employees who had shares made a great deal of money, while all the others shared a special bonus of $18 million, the equivalent on average of six months' pay. At the time, Celtel had 4000 employees, of which 98% were African."

But Mo Ibrahim did not keep the balance of the shares in MSI: in 1996, it merged with a specialist American software company and there was a share swap, diluting his shareholding as the founder of the company.

Celtel was born directly out of MSI. As part of its work of setting up new networks, the company set up a subsidiary called MSI Cellular Investments. In return for work it did, the company would be given shares as part-payment for work done. Over the years, the company acquired minority shareholding interests in places as far apart as India, Hong Kong, Uganda and Egypt. The last two were in partnership with Vodafone.

The spur to starting the new company came in 1996 when the venture capital company General Atlantic Partners took a 20% stake in the software company. As a condition of the deal, it wanted to demerge the MSI Cellular Investments subsidiary: its view was that investments in mobile companies did not fit with the rest of the business because of the much greater capital required. Terry Rhodes was in charge of the cellular investment business: "I used to wait my turn during the board meeting discussions on software only to be told I was mad and no one would make money in Africa."

So MSI Cellular Investments (later renamed Celtel International) was set up as a separate company in the Netherlands. The reasons for setting up there were mixed but proved to be far-sighted. As the manager responsible for the incorporation recalls: "This was really the tax tail wagging the corporate dog." But later it proved to be a good choice both in tax terms and as a place for raising corporate finance, as well as providing a very good travel base from Schiphol airport.

The company started life in a modest, serviced office in Hoofddorp, a small village not far from Amsterdam's Schiphol airport that was in those days surrounded by rustic Dutch meadows with grazing cows. Now it is a business hub full of shiny corporate office buildings (Celtel's own building among them), and a short commute from both the city and the airport.

Those recruited were often confused about the size of the company they wanted to work for. As one early recruit remembers: "I had no idea of the size of the company when I came in for an interview. It was a Regus serviced office space. There was a big meeting room for the interview. I started the following Monday at MSI so I turned up and asked where is my office? I was told you're sitting in this room and we've got another room next door. There were only four people and I thought there was a whole floor. But by 1999 we had moved to another building and had an entire floor."

Another early recruit found himself starting the job on the day he was interviewed: "I was in France and I read an article in a newspaper one day about 'MSI, a giant of African markets.' I sent a job application to the company. I was called for an interview on 7 July 1998, a date I remember to this day. After an hour of being interviewed, someone said: when can you begin? I said I'm available immediately so they said you can begin immediately. There was a delegation from the ministry in Chad there that afternoon so I met with them."

This new company, which would over time turn into Celtel, was Mo Ibrahim's next challenge. It was spun out of the consulting company providing expertise to operators but in 1998 it became a mobile operator. Although Celtel firmly established itself as a pan-African operator, it was by no means certain at this stage which way things would go.

But Mo Ibrahim was sure that there was a market to be found in Africa: "I was convinced that there was a market and that it would be a success but I had no idea to what extent. The need of individuals to communicate with each other is the same in Africa as anywhere else, but Africans felt frustrated due to the almost complete lack of available landlines. We were the first to take the bet that cellular telephony could fill this void. Some of the first networks were introduced in Uganda and Zambia and we immediately sought partners to invest alongside us. During the first five years, we made numerous financial presentations to potential investors."

Others were less convinced at the time. In the process of trying to encourage investment for the fledgling operation in Uganda, the company's executives had visited a well-known American telephone company. Having listened to the pitch, the American reached over to a hard hat he kept in his office, put it on and said: "I'm not going to Uganda without this. That's where Idi Amin lives." With as much good humour as they could muster, the visitors pointed out to the manager that Idi Amin had fled Uganda some 10 years previously. Without missing a beat, he pointed out that if he didn't know this, then none of his fellow board members at head office would know it either.

Apart from South Africa, at this point there was very little external investment in Africa because of the fear of political chaos. Mo Ibrahim, who knew the continent, felt that these fears were exaggerated: "Since I'm African by origin myself, I was obviously well aware of the daily difficulties faced by people in Africa. But I felt that my contacts greatly overexaggerated the risks, that there was an enormous gap between their perceptions and the reality. Through discussing and arguing for this idea, I decided that MSI Cellular Investments would become a mobile telephony operator in Africa."

Although things were to go in very different directions geographically, MSI started with a very interesting hand. It had minority stakes in operations in Uganda, India and Hong Kong and the right to acquire a 7.5% stake in an Egyptian mobile bid called Click GSM that became Vodafone Egypt. It was also pursuing licences in Zambia and Malawi.

Mo Ibrahim was a man who never went to business school and like so many who set up tech businesses he was an engineer, steeped in the mysteries of signal interference and network planning. He was not self-educated like so many entrepreneurs, who choose to miss out on learning because they are itching for the action a life in business offers. He got his PhD and, compared to many, spent a great deal of time on his education.

But even with hindsight, it is hard to explain how the son of a humble Nubian clerk came to Britain, made a successful career for himself in a BT company and then went from a grubby office in the Elephant and Castle to launch not one but two successful companies. Each of these steps is extraordinary by itself but taken together they say something about globalisation and opportunities.

For as one of the senior managers who worked closely with him in Celtel observed: "In the UK he was given opportunities not available in Sudan and Egypt, for example through privatisation." However, if Mo Ibrahim had stayed in Egypt or Sudan, he might have become a senior manager in a government-owned telco but he would never have created the wealth and jobs he was able to across Africa.

Individuals can succeed on their own merits in African countries but it is significantly harder without friends, influence and family connections. Even now, a number of key telecoms businesses in both Egypt and Sudan are either government owned or government influenced.

What Mo Ibrahim experienced in the UK was the way in which a society opened up opportunities, even to those who could be described as outsiders. Whatever the barriers racism threw up in the 1960s and 1970s, by the 1980s it was these outsiders who had begun to storm the gates and make a mark on the society that they had made their home. Perhaps if Mo Ibrahim had arrived at an earlier point, he might not have found the current of social mobility running as strongly as it was when he chose to build his first business.

One of Mo Ibrahim's talents as a leader was to bring people together and create a committed team. As the deputy chair of the Celtel board Lord Cairns observed: "Mo was extremely clever in putting together a team of people who shared his vision regardless of colour, creed or race. He would use talent wherever it came from, mixing nationalities liberally to achieve his goals. Sometimes it didn't work but more often than not, surprising choices made for extraordinary results."

Mo Ibrahim had learnt the importance of having a strong board which was one of the first things he assembled when he launched Celtel: "We needed a big board to show the kind of corporate governance that would give us credibility with the countries we dealt with and the banks. Also we needed to give comfort to the telecoms authorities. These needed to be respectable people. When I started I had 69% of the company. But I only had one seat on the company and there were 12 big beasts."

Also because he came to know and respect certain British values, he made these a central part of his business philosophy and ethos. His sense of fair play meant that he had a principled but nonetheless pragmatic view that making corrupt payments was not a useful business practice. Doing this in Africa meant negotiating fine lines, but he inspired his team of managers to be able to do the business without crossing the line.

By the time Mo Ibrahim had sold his shares in Celtel, he was to become a very rich man for the second time in his life. The UK *Sunday Times'* 2007 ranking of individual fortunes featured Dr Mohamed Ibrahim in 183rd place with an estimated net worth of £343 million.

CHAPTER 3

From corporate governance to political governance – Mo Ibrahim's Foundation

Mo Ibrahim is not the sort of man to let the grass grow under his feet. He has set up a US$100 million fund to take equity stakes in innovative African companies operating outside the telecoms field. Such a move might be expected of an entrepreneurial individual looking for the next opportunity. But setting up his own Foundation to address governance issues in Africa was perhaps more of a surprise, although not to those who knew that Mo had been worrying for years about the best way to improve political governance in Africa.

The Mo Ibrahim Foundation offers the African leadership prize. Mo felt driven to set this up because he understood that poor leadership or governance was one of the strongest forces creating poverty and underdevelopment on the continent and that to change the political culture requires starting at the top. The prize for the winning leader is a whopping US$5 million spread over 10 years, plus $200000 a year added for life for African leaders who rule well and hand over power to a democratically elected leader at the end of their tenure. Leaders eligible are those who have left office in the last three years. The award is higher than the US$1.4 million offered by a Nobel Prize Committee to winners of the coveted prize.

Mo does not choose the winner. That is done by a Prize Committee, the first of which was chaired by former United Nations Secretary-General Kofi Annan and comprised: Martti Ahtisaari, former UN Special Representative for Namibia and former President of Finland; Aïcha Bah Diallo, former Minister of Education in Guinea and Special Adviser to the Director-General of UNESCO; Ngozi Okonjo-Iweala, Managing Director, World Bank; Mary Robinson, former President of Ireland and former UN High Commissioner for Human Rights (and board member of the Foundation); and Salim Ahmed Salim, former Prime Minister of Tanzania and former Secretary-General of the Organisation of African Unity (and board member of the Foundation). After the Committee has reached its decision on the winner of the prize, there is a joint meeting with the board of the Foundation to approve the decision before it is publicly announced.

The first winner of this new award in 2007 was the former President of Mozambique Joaquim Chissano. By negotiation Chissano had ended an extremely bitter civil war that had lasted 16 years and brought his opponents into government with him. He took office after the death of Samora Machel in a plane crash in 1986. The civil war had devastated Mozambique, left thousands dead and forced many to flee their homes. Chissano took on the task of governing a country whose infrastructure and economy were ruined, its society deeply divided and which suffered from severe natural disasters.

Chissano was elected president in multi-party elections in October 1994 and again in December 1999, announcing that he would step down from office in 2004. He refused a third term in office as president and decided that as he was 68, it was time to make way for a younger person.

As Chissano said in his acceptance speech at the prize-giving ceremony: "My decision was largely influenced by the understanding that the country was in peace and the economy was steadily growing. Democracy was taking root. I realised the time had come and conditions were right to allow a new leadership to take over and push the country forward." Although no longer in government, Chissano launched the Joaquim Chissano Foundation in November 2005 to promote peace and social, economic and cultural development in Mozambique.

In Africa this was exceptional behaviour, as many long-standing leaders look for a lifetime in power and are only really dislodged (by election, coup or popular revolt) against their will. As a result, some African leaders are extremely old and often out of touch with ideas and technologies that have developed over the last decade or two.

However, once the award was agreed, the chair of the Committee Kofi Annan tried calling Chissano to tell him he had won the award. He was unable to reach him on any of the numbers he had provided. Inevitably the question asked by journalists at the press conference for the announcement was: what was Chissano's reaction on hearing that he had won the prize?

Eventually Kofi Annan managed to contact someone, who told him that Chissano was somewhere between northern Uganda and southern Sudan, trying to speak to the Lord's Resistance Army about peace negotiations that were taking place. Annan told the bewildered person who had answered the phone: "Say happy birthday to him from me." It is perhaps a measure of the man that he had chosen to be brokering peace negotiations rather than celebrating his birthday. A genuinely surprised Chissano was eventually reached by the BBC and when told of the size of the financial award seemed for a moment rather overawed by the whole thing.

The prize is striking, but at least as important is the Ibrahim Index of African Governance, a new, comprehensive ranking of sub-Saharan African nations according to governance quality. This assesses national progress in the following five areas: sustainable economic development; human development (health and education); transparency and empowerment of civil society; democracy and human rights; and rule of law and security.

It has been developed under the direction of the Kennedy School of Government at Harvard University. Known as the Ibrahim Index, and first published in September 2007, Mo hopes that its regular publication will lead to an improvement in the way in which the citizens of sub-Saharan African countries are governed, and stimulate debate about the criteria which comprise good governance.

As Ibrahim sees it: "We want to give African civil society a tool to assess the performance of their governments in an objective way. Every criteria used has been referenced and so is difficult to dispute. There are 58 measures in all that capture the key elements of government performance. So over a number of years a clear picture will emerge as to what progress has been made and to see how countries move up and down the scale." Not surprisingly, countries with armed conflicts are all at the bottom of the Index. Ibrahim's own place of birth, Sudan, has a low placing because of the conflict in Darfur.

The Index touches on a topic that is sensitive in Africa: how can a continent that has been so blessed with such a wide range of natural resources be at the bottom of any objective measure of global poverty? Ibrahim is clear that "it is not enough to be blessed with natural resources; you need really good governance to marshal these resources, to increase the human resources and to instil a sense of citizenship and fairness."

At the formal ceremony in Alexandria the award's chairman Kofi Annan handed the Prize to Chissano and highlighted the importance of leadership in this process: "The Prize celebrates more than just good governance. It celebrates leadership. The ability to formulate a vision and to convince others of that vision; and the skill of giving courage to society to accept difficult changes in order to make possible a longer-term aspiration for a better, fairer future."

At a World Economic Forum meeting in South Africa in 2007, the participants were called upon to celebrate the 50th anniversary of Ghana's independence. Speaking on one of the event's panels, Ibrahim asked "what exactly happened since independence?" In Ghana's early years of independence, it had a higher standard of living than places like Malaysia and Taiwan. South Africa's President Thabo Mbeki responded by saying that surely Ibrahim was not suggesting that colonialism was a good thing. In his usual forthright way Ibrahim replied by saying that we made a mess of post-colonialisation and we need to be clear why we made that mess.

Reflecting on this question later, Ibrahim believes that "many of the great fighters for our independence did not have the right make-up to be sensible post-independence leaders. The duty of nation building is different from fighting for independence. We tend to worship the leaders who brought independence but although some of these people were good, they were not necessarily good leaders. Charismatic freedom fighters did not allow the growth of institutions. Criticism of them and freedom of speech were stifled."

Issues of post-colonial borders weigh heavily in the balance. The borders of Africa's new nations "cut across ethnic groups and economic zones irrationally and became the seed of endless ethnic conflicts." The Cold War meant that in a choice between two evils, the Western nations tended to support leaders on the basis of a tried and tested principle: he may be a son-of-a-bitch but he's our son-of-a-bitch. Leaders like the then Zaire's Mobutu were allowed to get away with behaviour by institutions like the World Bank that would be cause for criticism and action once the Cold War had ended.

"Guys like (Nigeria's former dictator) Abacha are psychopaths. But what about the British bankers who accepted the sums of money he sent out of his country? There has been failure on all sides and proper behaviour is needed at an international level. The difficulty with this type of leader is that they are megalomaniacs who define themselves as a country and don't see the difference between their pocket and the Treasury.

"Corruption is a problem that you need to deal with before it becomes part of the culture. Mobutu used to say to his own people: go and raise your own funds and the message was clear. But things have changed. For example, in Kenya the international community objected and said you can no longer carry on with this kind of monkey business."

And whilst business has often been a willing partner in this "monkey business," Ibrahim believes that based on his own experience, it is in the private sector's interest to have an environment that is clean of corruption: "For business creates jobs. Well paid jobs create a middle class. This middle class creates knowledge, ethics and culture."

The role of the media and increased communications is hugely important as one key element in these changes: "When large numbers of people have a mobile phone, everybody knows what is going on. They are able to communicate what is happening to others and it is no longer possible to decide things in a backroom. Connected people are difficult to deceive."

CHAPTER 4

The talk business – how does it work?

n the years before mobiles in Africa, communication for the average citizen was extremely difficult. In many cases you had to walk to talk face to face. The postal system was more or less non-existent. The government-owned monopoly telecoms companies ran their business on state planning lines that bore no relationship to market demand. The capital costs of fixed line network expansion were high but were owned by governments, typically short of finance, who starved their telcos. Vandals frequently stole the copper cable network that was already in place. Civil wars destroyed whole networks in a significant minority of countries. With only a few notable exceptions, most incumbent telcos were lethargic, staffed by political appointees and had little or no idea of customer service.

Like much else in Africa still, the telecoms incumbents were selling shortage: there wasn't much capacity, it cost a lot of money and if you didn't get a very good service, there was no point in complaining. Prospective customers might remain two to 10 years or more on a waiting list. The knowledge that the waiting list was so long obviously deterred all but the most determined, and this fed the interests of those who could arrange that you jumped to the top of the list for a fee, or a favour.

Government was the biggest customer of the telco incumbents and out of this close relationship a vicious downward spiral was born. Government ministries and agencies did not pay their bills or paid them months or years later. As a result, the telcos were starved of cash, and because the telcos weren't collecting revenues, they lacked the capital to invest in new equipment. As a substitute, tied-aid was often used and therefore the companies built up a patchwork quilt of incompatible equipment.

As a large state-owned agency, the telecoms companies were also regarded by many governments as vehicles for patronage. Jobs were often distributed to political supporters and revenues were on occasions siphoned off to support the election activities of ruling parties. The combination of civil service procedure and patronage slowed the pace of development to a standstill. Some telcos had multiple layers of management that no longer reflected functions or any sense of the task in hand.

The arrival of mobiles was like a clap of thunder in this slow-moving pantomime. As Terry Rhodes recalls: "The village and the city can be a two-day bus ride from each other. People would send scraps of paper with messages on them to each other by bus passengers. Or they'd go themselves. So we were competing with a two-day bus journey and were only charging US25 cents a minute. It really was about making life better in Africa."

Before mobile phones, people were employed as runners to take a message from one side of the city to the other, which would involve multiple journeys. Again Terry Rhodes remembers these strange times: "It was like Roman times, when letters were sent by runners. It was more walk, less talk. In Zambia, you had to book a time for an international call. So you would say that you wanted to book a call to Europe at 8pm tonight, and the operator called you when there was a circuit free."

So not only was Africa having difficulty meeting the two first most basic needs – food and water – it could also not meet the need to communicate. In contrast to this lack of supply, there was an enormous pent-up demand that nobody could really calculate. African societies and cultures are predominantly oral; people talk a lot.

Whilst some part of this might be put down to low literacy levels, there's something in the DNA of most Africans that makes them almost Olympic standard talkers. For example, the idea of a "quiet Nigerian" is almost an oxymoron as Nigerians themselves will cheerfully admit. But at another level, Africa is no different from anywhere else in the world today: if you walk the streets of the big cities, you will see people everywhere talking on mobile phones. The demand is seemingly endless.

The first of the three problems to be solved, as Telecel had recognised in the mid-1980s in DR Congo, was that of supply.

The business model for a mobile phone service is completely different from that of the fixed line business. It all comes down to how the customers are connected to the network – in industry jargon "the last mile."

Fixed line telephony is based on physical locations. There is a cost per line, pieces of copper wire are connected to a person's house and the operator probably charges some version of a monthly fee. Whether or not a person is using their phone, this fixed line connection is costing the operator money, in terms of maintenance and interest paid on the money invested to build it.

By contrast, a mobile operators' "last mile" is built according to the traffic handled – the operator does not incur any customer-related costs when the customer is not using their mobile and thus there is no traffic.

In relative terms, it hardly costs the operator anything. No heavy costs are triggered until you make, or receive, a call.

The cost which is then triggered is the cost of the radio equipment (the transmitters/receivers or "TRX") needed to transmit the call, along with the towers needed to hold the aerials up in the air, the power needed to operate the radio equipment and the air conditioning often needed to keep it operating in the African climate.

Since a normal subscriber uses their phone for maybe 10 minutes a day on average a fixed network is generally idle for the other 23 hours and 50 minutes – so the cost of handling an incoming call is negligible. However, when a mobile network handles an incoming call from another network it has to use additional capacity and it has to search for the customer who could be anywhere in the country. Thus the mobile networks ask for, and regulators usually grant, significantly higher fees for terminating a call onto mobile networks than onto fixed networks.

However, in terms of capital costs these roll-out costs are far kinder to the mobile operator than the fixed line operator. The mobile can build its installations as the market grows. It only has to switch on a handful of base stations initially, and can add in more base stations once there is sufficient traffic to justify it. The initial capital expenditure can be only a few million dollars. Thereafter it can expand whilst the business is running.

Although the advent of the GSM standard in the mid-1990s helped to solve the capital expenditure problems by producing equipment and phones on a global scale at ever-decreasing cost, it was not yet enough for Africa. The traditional model of making calls first, receiving a bill at the end of the month and then paying was always going to be difficult in Africa and it meant that at first only a small elite could participate.

But the potential demand went well beyond the elite. Phones were about something more than simply the functional need for communication. They represented a sense of aspiration for people with very little who had few opportunities to show they wanted better. Phones were sexy and high status in every country. If you were an African government minister, you did not want to admit to your peers that your country did not have a mobile network.

When pre-pay systems started to appear in Europe in the late 1990s the second part of the puzzle was in place. The operators could stop worrying about credit control and be assured that they would be paid for their usage. By 2001 the vast majority of customers on African networks were pre-pay: in other words, they paid up-front for their mobile service. Not only that, if there were to be millions of them, they needed to be able to buy their service as easily as possible. Users needed to be able to find SIMs to activate their phones and scratch cards to buy minutes everywhere.

In 1999 MSI had won a number of licences across the continent and needed to raise finance in order to purchase the capital equipment to build the initial networks; to do this they had to approach a number of bankers and other finance institutions with a business plan. This had to be a good, believable, story.

A question that was seemingly always asked was: "Fixed telephony has been in existence in Africa for 100 years, why do you think you can build a viable business?" MSI's answer was based first on the recognition of the political and economic limitations of fixed networks, and second on the advantages of mobile. The GSM radio technology had matured and was now significantly cheaper than it had been for initial deployment in Europe. The pre-pay systems such as those which MSI introduced in Uganda solved the cash collection risk and increased the potential market enormously.

But the challenge, as the then CTO of Celtel Moez Daya observed, was "to build the best networks we could in a tight budget … We didn't (initially) pay for post-paid billing and voicemail systems. These could have cost US$5–10 million. The systems we did buy were good enough for the job and kept the quality high but were basic and simple."

The company started a planning board that looked at and approved the return on capital investments at all levels. And it was able to negotiate hard with equipment vendors to get the best deal possible for the company. As Daya remembers it: "We started out with Siemens and then Alcatel and Ericsson came around. In Tanzania we decided to use Alcatel, partly because of the vendor financing offered.

"Vendors tended to sell you their products on the basis of subscriber numbers but we managed to get them to accept a payment by minutes (of calls made). They saw what we were trying to build and as a result, our costs were lower than our competitors.'"

Like most operators, Celtel was always trying to work out whether it was best to have two competing equipment suppliers or one who would offer large discounts for being the only supplier: "Ericsson wanted to become the common core supplier ... We agreed and in this role, it replaced all our old systems with the latest technology at almost no cost."

As far as pre-paid was concerned the operators quickly discovered that it was always the lowest denomination scratch card for minutes that sold best. This made it an everyday consumer purchase like cigarettes, chewing gum and Coca-Cola – in business jargon a "fast moving consumer good." So you're trying to make money on a high frequency of low-value purchases. Luckily for the mobile operators, in Africa there was already what Terry Rhodes described as "a beautiful informal distribution system for all of these things and all you're doing is adding another thing in."

But in some ways the speed of turnover of scratch cards made it more like newspapers because people wanted them every day. Some distributors would come in and buy $1000 worth of scratch cards in the morning and be back in the afternoon for another $1000 worth. This was like magic to them because they didn't have $2000 in the morning. Today the distribution margin across all these distributors is worth more than $3 billion through-out all African countries. The effects were bound to be hugely significant: a whole new form of economic activity had been invented with all the jobs and spin-off effects that come with new areas of economic activity.

Nowadays all the operators work through a combination of dealers (who in turn deal with street sellers), their own shops and, recently, elec-tronic distribution of airtime. Rick Beveridge identified the reasons for Celtel's original choice being that "we decided that people who worked for themselves were more likely to do better than people working for us. Not only were they more motivated but we trusted the market to find good solutions to the distribution problem that we would never be able to do ourselves." The distribution channel takes 10–12% of overall revenues. Increasingly the commission structure is focused on recruiting long-term, higher value customers. Today Celtel has over 500000 retail outlets across Africa with flagship retail outlets to serve marketing and branding purposes.

Because most people's disposable income is limited in Africa, some of these effects were unexpected. The cigarette vendors, soft drinks breweries and even petrol distributors all complained that their sales were going down as disposable income was spent on mobiles. Indeed, in the early days a group of Ugandan MPs representing brewing interests in their country even proposed that mobile airtime sales be limited to protect beer sales. Interestingly today Ugandans spend more on mobile airtime than on food.

Governments were not slow to spot this new gusher of economic activity. Soon the operators were hit by a succession of taxes that quickly totalled 30–35% of the overall cost of a call. Mobile companies are now one of the largest parts of the tax base for almost all African governments. Sadly high tax levels keep mobile calling rates higher than they might be and thus deprive poorer customers of the opportunity to use them.

So who are these customers? They can be almost anyone, although the majority are found in urban areas. This is an African market that nearly everyone is a part of. Even where someone cannot afford a phone, they often share someone else's phone. A survey of the inhabitants of Nairobi's massive slum Kibera found that each phone was shared by four people.

More than 90% of the revenue from customers comes from domestic calls, the rest coming from international calls and SMS. Because of high levels of illiteracy in many countries, the mobile operators set up call centres: people often cannot read written instructions. For the first time 24-hour call centres appeared in Africa. Because of the nature of the business it was important to form relationships with customers if you wanted to hang on to their business. This was in stark contrast to the high-handed behaviour of the telecoms companies before the arrival of mobiles.

In Celtel's case more than 99% of customers are pre-paid. Every effort is made to reduce the cost of entry. There are limited handset subsidies but in some countries there are affordable handsets as a substitute. The plentiful availability of second-hand mobile phones offers an alternative. There are continual promotional offers on SIM cards that year by year seem to get cheaper and cheaper. All of this enables the operators to get a shorter breakeven time.

A further innovation was a recognition that a large proportion of African consumers live from day to day – they cannot afford to buy in large quantities and tie up cash filling a car with fuel or stocking a fridge with beer. A regular phenomenon is that when you take a long taxi ride, say from central Dar-es Salaam to its airport, the driver's first action is to pull into a petrol station and ask you for an advance on the fare by putting some fuel in the car; he would not normally have enough fuel to go that distance.

In scratch cards this has meant that affordability is increased by reducing the denomination of the cards – a "bottom of the pyramid" customer who may not be able to top up $5 once a month may be able to top up $2 three times or $1 seven times.

The latest innovation is the ability to transfer minutes electronically thus removing the need to produce plastic scratch cards. It also lowers the transaction cost and thus will make even lower denomination sale values, thus enabling the operators to enfranchise yet another section of users into the market.

Although not always an accurate guide to profitability, average revenue per user per month (ARPU) is the benchmark used by the industry. Although these vary widely, they range between US$8–40 across the continent. In India, some operators are operating profitably on ARPUs of between US$4–7 so the bottom of the price elasticity curve has not yet been reached.

In 1999–2000 when the first pre-paid operators started appearing, despite the close similarity to fast moving consumer goods (FMCG), no one paid much attention to marketing. The operators, Celtel included, were dominated by engineers and financiers. Although the market in Europe had already surpassed 50% penetration it had largely done so on the back of post-paid, contract sales and subsidised handsets. The industry was dominated by building networks at considerable speed and most senior managers came with a technical background. Even in Europe hardly anyone in the mobile phone industry paid much attention to marketing, with only "Orange" having ever built a major consumer brand. So the overall expenditure on marketing in the early days of Celtel was an almost unbelievably low 2–3%. With a far greater level of competition, this has risen to around 8% excluding the costs of add-ons like loyalty programmes.

Another major part of the cost per minute is the need to pay other networks for interconnection, costing somewhere around 19–20% of the overall cost of a call. Interconnection is a fraught topic in the industry for a number of reasons. Initially, the fixed telecoms companies refused to connect other operators to their networks, thinking that they had a right to hang on to their potential customers despite patently failing to deliver any service whatsoever to them. After that hurdle had been overcome, the fixed operators in particular, for example in Uganda and Zambia, often complained about the level of interconnect charges as they found themselves delivering more and more calls to and from the new mobile networks. Worse still, they were often very reluctant to pay their interconnect bills.

So what do customers make of the service they get? In one of the customer satisfaction surveys used publicly after its rebranding exercise, it showed Celtel achieving a 70% satisfaction index and increasing its market share from 42 to 48% after the rebranding had been completed.

Indeed, once a mobile operator has built the basic network of towers to cover a city or geographical area it costs almost nothing to acquire each new customer, and very little to add new capacity to service new customers, a similarity with another aspect of the FMCG business, the supermarket. Therefore there is a considerable incentive for operators to "pile it high and sell it cheap," in the words of Rick Beveridge. Since 65% of capital costs (capex in the jargon) is the purchase of radio transmission capacity, which is constantly being rolled out into new geography, there is an overwhelming need to maximise revenues from it. Due to the problems with the fixed operators there is a wish to get as much of the traffic as possible on your own network by encouraging your customers to call each other.

In order to create a mobile network, there are three pieces of technology: a radio network to connect to the customers' handsets; a transmission network to link the local base stations; and a network core consisting of switches, billing platforms and equipment to track customers geographically – all of which is necessary in order to route calls and charge the customers for making them.

In Africa there is a fourth piece: a power network. Africa has very little power infrastructure: a satellite picture of Africa at night shows an almost entirely dark continent except for pin pricks and slashes where its urban conurbations can be found. By contrast, large parts of the USA and Europe are lit very brightly because of far higher levels of urbanisation.

But some places in Africa have hardly any infrastructure at all as the then Celtel manager in Gabon Emily Macauley discovered when visiting a new location to instal a base station: "I once had to take a plane, a helicopter, a boat and a canoe to reach my destination. There was a small port with illegal, informal trading. The need for telecoms was very high, everyone from the secret service who needed to control the traders to the traders themselves. (Once it was installed), we had to double capacity in less than a year." A country like the Democratic Republic of Congo (DRC), which is the size of Western Europe, not only has no power but very few roads.

In coverage terms, most African countries have now reached a point where 65–70% of the population of a country has been covered. In the next three years this percentage will probably rise to 80%. Coverage is obviously easier in geographically compact countries. Satellite data showing population density and "on-the-ground" knowledge of local spending patterns is used to plan networks.

Emily Macauley's experience in Burkina Faso demonstrated to both her and Celtel that getting out into the most unlikely places produced more growth: "You have to spread the customer care out to places like the desert in Burkina Faso. We opened regional offices in the country and on the basis of these we had 100% growth in a year."

Because of the absence of power, each base station has often to be powered by two generators, one for regular operations and one for back-up. Three years ago one operator in Nigeria had over 5000 generators to look after. No wonder one of its managers joked: "I don't know whether I'm in the power or telephony business."

If anything goes wrong, it won't just take hours but literally days in some cases to get there to fix it. Forty per cent of Celtel's employees work in network operations. A base station might cost the operator in the region of $250000 for everything: the mast, the generators, the equipment to transmit, air conditioners to cool it, fencing and a hut for the equipment, even more if, as in some countries, the government imposes import duties of up to 40%.

Luckily many local people have worked out that most of the equipment is valueless to anyone other than the operator so base stations do not suffer the vandalism experienced by fixed line operators – so long as you protect the one thing that does have value to others (the diesel for the generators), which is usually protected by the presence of a security guard, who often lives alongside the base station and cultivates the surrounding land.

Base stations have to go somewhere and buying land is often challenging as there may be no clearly established ownerships or land registries. Two people may have extremely credible pieces of paper claiming that they own the same land but only one of them actually does. In the larger cities, the very act of acquiring sufficient land for base stations has driven up real estate prices.

Mamadou Kolade described the process of acquiring sites in Congo-Brazzaville (Congo B): "We used the office tower owned by Hydro Congo, the state oil company. Then we went to the schools. You pay and get an agreement and give them books and materials. But there's always an issue of whether you're talking to the right person. We made an agreement with a national government organisation and paid a local guy in Pointe Noire for the agreement. The big chief in Brazzaville didn't get any of the money and came to complain. So we showed him what we had paid his local guy."

A great deal is spent both on fuel and on transporting it around the country. A typical base station might cost $2500 a month to run but in an out-of-the-way rural area this figure can go as high as $20000 a month. In one of the more extreme cases on the continent, an operator has to send the fuel by boat where it is then hand-loaded on to a lorry for the final portion of its journey. Using trucks on roads in the West African rainy season very rapidly wrecks them beyond the point where maintenance is essential.

When GSM technology was first used in Europe, it was much more expensive because there were fewer operators and fewer users. Now, as a mature technology, it is cheap, readily available and "off-the-shelf," and at this stage it could almost have been designed for Africa. For it is a capital intensive business and 20% of costs come from capital depreciation.

Rick Beveridge quantified the impact that the fall in equipment costs had on the bottom line: "When we started there had been five years' experience of GSM technology and 100 million customers worldwide. There were only a handful of vendors: Alcatel, Ericsson, Motorola, Nokia and Siemens. By the time we were setting up networks in Africa, the vendors had bought the prices down. The equipment used to cost $1000 per customer in the UK but by now this had come down to $100–150 in Africa."

All of this activity in Africa would be exciting in its own terms. It showed that it was possible to get things done at great speed and serve customers previously thought too poor to matter. But it has had a much wider impact on all of the countries of Africa. The process of setting up networks has passed like an electric current through the sometimes sleepy cultures it has affected. The mobile operators have quickly become major employers, often paying top-of-the-range wages to all categories of employees.

Whilst a few management posts in all operations have been filled by external managers from outside the continent, the operators between them have created the largest cadre of Africa management experience. Unlike many of the companies that dig or pump things out of the ground, they have promoted staff from within countries and put a wide range of skills into the local economy.

A typical example of this process is the career path of Cape Verdean Emily Macauley. Being a woman manager is unusual in itself but she passed through commercial positions in Burkina Faso and Gabon before being promoted to overall country manager in Burkina Faso and then Madagascar. The sheer scale of recruitment benefited African countries enormously. At one stage the company was recruiting 250 international level people and many more local Africans.

But there were also ripples out into the wider economy. Emily Macauley's favourite memory illustrates why: "In Burkina Faso, I saw a welder in the street. I said to him, if I give you a picture of a kiosk, can you build it? At that point we were importing them in fibre glass from South Africa. He built five different models until he got it right. I then asked him to do 20, and then 100. One year later he had done 150 kiosks.

"I presented what he had done at a Celtel managers' conference and he started getting work from our other subsidiaries. He went to Niger and Chad. When I met him at the airport once after this, he was in a suit and had negotiated a patent on the design. He said: 'I want to thank you. I'm building a home, I've sent a child to Europe for schooling and I now have 300 people working for me.'"

This clearly demonstrates that Africa has the potential to generate wealth that can change lives and does not need to be the donor-funded, wounded, whimpering animal of the world for the rest of time. But this wealth also washes over into social responsibility as the company has a very active corporate social responsibility programme. In Burkina Faso,

Emily Macauley set up a scholarship scheme to educate young girls abandoned in rural villages. This same scheme has now been extended to all youngsters in remote villages.

But the description of the mobile business in Africa above was hard wrought from experience and it all could have turned out very differently, as Celtel's experience in Uganda showed.

CHAPTER 5

Starting out is hard to do

MSI's first opportunity to run its own mobile operation came in Uganda in the mid-1990s. It had met a group of Ugandan entrepreneurs who wanted to open a mobile network in their country and they sought MSI's help. It decided to take the project to Vodafone.

MSI met Vodafone's International Business Development Director, Julian Horn-Smith (now Sir Julian). As he had previously worked at the confectionary company Mars, he would sometimes jokingly refer to himself as "the man from Mars." Other than planning its involvement in South Africa, Vodafone had not shown much interest in Africa up to then. However, on this occasion, having been persuaded by MSI, it decided to dip its toe in the water.

MSI was quite modest about its own role at this stage because it did not think of itself as a fully fledged mobile operator. Founder Mo Ibrahim was more interested in securing the trust of Vodafone: "We went in as the junior partner. We didn't want to be the lead operator because at that time we were selling software and consultancy. Vodafone were appointed as project director and we deferred to them. At this stage, their top management was involved: for example, Chris Gent (now Sir Chris) was the chairman of the board so he had to turn up for board meetings in Uganda."

One of the reasons why Uganda managed to attract investment of this kind was because it was one of the first countries in sub-Saharan Africa to open up its economy.

For as Mo Ibrahim told Uganda's *The Monitor* in an interview in 2004: "In 1994, Uganda was much more open to foreign investment and liberalisation of the economy. (It was) ahead of other countries at that time. That is why we had a cellular industry there before Kenya and Tanzania. The tremendous goodwill from the international community towards Uganda was another vital factor for us to come to Uganda. When we came here, we did not have to bribe anybody. It was open. And up to this point, other countries saw foreign investors as people who have come to steal things from the locals."

As a result of this liberalisation, total investment more than doubled from $117 million in 1996 to $295 million in 1998 and part of this investment was from mobile operators who would continue to increase investment in the country over the next decade as they rolled out their networks.

The Vodafone–MSI launch strategy reflected prevailing thinking in the mid-1990s about how the mobile market in Uganda would evolve. The expectation was that they would be luxury items for the country's elite and coverage would be mainly in the capital city with perhaps additional coverage in one or two other cities. As one of those involved remembers: "We were charging a premium price and reaching profitability quickly. We were recording a profit by the second year. We only discovered the huge market potential much later on."

High prices meant small numbers of subscribers, as they were then known long before they became "customers." The network was launched in 1995 but three years later it had only recruited 5000 subscribers. The lucky 5000th subscriber – Flavia Muntuyera – was given 5000 free minutes. Nevertheless all should have been well because for the moment Celtel (as the network was branded) was the monopoly mobile operator and high prices meant it had good, quick returns on its investment.

However, the key weakness was having to follow the fixed line operator model of allowing customers to pay a month in arrears, which was known as "post-paid." Furthermore, customer billing data needed to be transferred back to Vodafone's Newbury base in the UK before bills could be issued to customers, something that incurred delays often running into months. The result was that bills went out late and many customers showed a marked reluctance to pay them. Celtel Uganda may have reported profits, but collecting the cash was altogether a different story.

A steady stream of local press stories began to appear as non-paying clients were taken court. A typical report of this kind appeared in 1997 in *New Vision* when the company took 82 individuals and firms to court who collectively owed the company $64308, an average of $784 per customer. Many had been sent reminders up to 10 times.

The company tried embarrassing non-payers by advertising their names in the local press prior to legal action. But neither public embarrassment nor legal action made much impact. By February 1998, it filed a suit in Mengo Chief Magistrates' Court pursuing 100 customers for just over $160000 in unpaid bills.

These were not Ugandans who couldn't pay but clearly people who (for whatever reason) had chosen not to pay, for the names of those taken to court read almost as a "who's who" of prominent members of the Ugandan elite: the Deputy Katikiiro of Buganda Grace Ssemakula; Premier Lottery, Gomba MP Israel Kayonde; a prominent tycoon Mohamed Majyambere; a director of Bakayimbira Dramactors, Charles Senkubuge, Major Ruranga Rubaramira and the UPDF's Major Charles Otema.

Not only were many subscribers not paying but the level of debts incurred by individual subscribers was rising. Eddy Sebaggala, brother of former Kampala mayor Nasser Sebaggala, was sued for failing to pay mobile phone bills totalling $10932 incurred over a year. Again he was only one of many subscribers taken to court.

In 1998, when MTN arrived to compete with Celtel, the company had doubled its subscriber base to 10000. But perceptions of its reputation were not good for, as a report in the *East African* noted just before MTN's launch: "... many of (Celtel's customers) were unhappy. The system was overloaded and its clients griped that its customer service people were rude and arrogant." Whether true or not, the company now had the worst of both worlds: significant numbers of its customers were not paying and even those that had were not entirely happy with the service they were receiving.

Worse still, it was about to enter into a dispute with the Ugandan incumbent fixed line operator over interconnection costs that would bring it into conflict with the government itself. Interconnect charges are what one phone company pays another for calls to its subscribers. With only one company these were not something anyone needed to pay much attention to but with the arrival of mobile competition for the incumbent, they became a running sore for the new competing mobile operators in Africa.

In 1993, the fixed line incumbent, known then as Uganda Posts and Telecommunications (UPTC), signed an agreement with Celtel for interconnection. The cost of a call from a UPTC phone to a Celtel mobile was Ush650 a minute compared with the fixed line network call charge of Ush75 a minute.

Globally, the basis for interconnection charging can be enormously complicated. But the justification for this disparity is very simple: Celtel had spent a large sum of money on its network and the fixed line callers, in paying to use this network, were in effect paying a price that reflected the costs of this investment.

But it might reasonably be asked, why did UPTC sign what it was later to describe as a "lopsided" agreement? The fixed line operator's management did not believe that the new upstart would outstrip its traffic levels, or that the balance of traffic would turn from fixed to mobile rather than from mobile to fixed. For as David Sserunjogi, Celtel Uganda's Company Secretary, noted: "In the first four months of our operation, traffic was in UPTC's favour and we paid them $91458 in October 1995. Problems started when traffic started changing in our favour and they could not pay."

He maintained that if UPTC paid what it owed then Celtel would be happy to renegotiate the agreement. But with the bill standing at over $1 million there was little sign that the incumbent would pay up. Indeed, in a precursor of disputes that would recur time and time again across the continent, it was maintaining that the agreement was so lopsided that it was subsidising Celtel. David Sserunjogi also said that he had asked UPTC to state which aspects of the agreement they are unhappy with. "But they have never got back to us!" For as Mo Ibrahim recalled, in terms of financial credit, this was a largely one-sided relationship: "Because if we bought leased line circuits from them (on which they had a monopoly) we had to pay promptly."

But UPTC's strategy was clearly to renegotiate the contract in its favour through a combination of withholding what it owed and political pressure from its owners, the government. John Nasasira, at the time the Works, Transport and Communications Minister, issued a comment that clearly favoured UPTC: "This is very bad. I am investigating the contract. Our people must be careful when they sign contracts. This is not the first time. I want to find out whether this contract can be renegotiated." Amid the name calling that followed this dispute, UPTC frequently referred to Celtel a monopoly: presumably it took one to know one.

With hindsight the balance of traffic was not the surprise it seemed, because of two effects. First, a call to a mobile phone is a call to a person whereas a call to a fixed line is a call to a place. When customers want to talk they generally want to talk to an individual so, given a choice, they will tend to call the person (the mobile number) first. Second, the proportion of incoming calls that are usually answered on a mobile network is usually much higher than on a fixed network. The combination of these effects means that the traffic from a fixed network to a mobile network is usually around four times the traffic in the other direction. However, in the mid-1990s this effect came as a surprise. On the other hand, the introduction of mobile phones had expanded the market and UPTC's revenues had increased – Celtel was only asking for its share of that extra revenue.

Against this backdrop of mixed customer feelings and a simmering political dispute about Celtel's interconnection agreement, MTN secured its Second National Operator licence, which included the ability to launch a mobile operation and a gateway licence to carry international calls. MTN's marketing manager Eric Van Veen gave notice of the company's intended strategy: "We will attract subscribers from across the spectrum from senior business people to the ordinary person. We are going to make the monthly cost of phone ownership go down, that's what we're aiming at." In October 1998 it launched its network with considerable razzmatazz and high levels of advertising in local media with its distinctive yellow and blue colours.

Suddenly, Celtel had a fight on its hands and it was a fight in which the rules were going to be completely different. Initially, MTN, as it had promised, lowered both access and calling charges. Celtel had been charging customers $2000 for a handset and access to its network with occasional tactical promotions offering $100 off. MTN immediately brought this down to $400. As the *East African* observed at the time: "Those who paid $2000 were thought to be in a vengeful mood."

MTN also introduced calling rates that shook up the cosy duopoly between Celtel and the incumbent (now called) Uganda Telecommunications Ltd (UTL). Celtel had been charging US49–66 cents a minute and UTL US16–33 cents a minute for fixed line calls. MTN came in and undercut both of them with rates about half the previous levels. Celtel was forced to respond by cutting its local call rates at between US23 and 29 cents a minute. A price war followed that opened up the market and really benefited a much wider number of consumers.

MTN said it was anticipating getting 2900 subscribers in its first few weeks but had in fact recruited 4900 subscribers in its first day alone. The mould had been broken and once Celtel had learnt this lesson, it was a pattern it was to experience in many other countries. By the end of the first week, MTN had 6000 subscribers and looked set to overtake the Celtel subscriber base that had taken three years to build up.

But the killer blow came in 1999 when MTN introduced the first prepaid calling cards: customers could buy "scratch" cards of a certain calling value and this would be loaded on to their phone once they entered the PIN number on the pre-paid card. Pre-pay was only just being introduced in Europe but it was the solution to the credit control problem that had plagued Celtel and was taken up by the market with alacrity.

Celtel quickly introduced its own pre-pay service. But it was wrong-footed and suffered from shortages of its own pre-paid cards. Customers were not happy: "The $10 top-up cards ran out of stock two weeks ago but these people (Celtel) kept on telling us that they will be in stock any time. Since then, we cannot see anything," said a customer. Another customer said the $20 and $50 cards also ran out on Sunday.

A year after its launch, MTN had 55000 subscribers against Celtel's 15000: the company had well and truly taken a beating. But the bewildered incumbent UTL also emerged blinded and blinking into this new world. Its number of fixed line customers grew extremely slowly and then eventually ground to a halt.

Competition in the market had revolutionised how customers felt about getting a phone. This had been a country where the process of getting a phone had long been complicated. An individual had to fill in three copies of a four page form with photos attached and then spend weeks travelling many kilometres getting signatures from local officials and employers before the state-owned company allocated a phone line. After that, it could take two more years and several bribes before the fixed phone was actually connected.

Now, these same people could walk into a mobile phone outlet, buy a handset and a SIM card, walk out and talk. There were no application forms to fill in, no interviews and no bribes needed to be made. As a local paper *New Vision* observed: "Technology and the competition between Celtel and MTN have given the people an extraordinary amount of freedom, something which many never thought they would see in their lifetime."

Vodafone took a tough line on the interconnection dispute payment, saying that it would not invest further until the matter was resolved. In effect, the company was put into a holding pattern, contrary to the direction that MSI wanted to follow. This was not the first disagreement between the two owners. MSI had wanted to go after the Second National Operator licence in 1997, which MTN ended up winning, but Vodafone held back.

Suddenly, the local Ugandan partners wanted to exit the company and Vodafone waived its pre-emption rates. MSI then bought these shares and ended up as the majority owner. A year later, with no sign of resolution on the interconnection dispute, Vodafone decided that it was a marginal investment and that it too would sell its shares to MSI.

Vodafone had bigger fish to fry elsewhere – by now MSI had the experience and skills to be a mobile operator and decided to hang on to the operation, even though it was loss-making. The mistakes were to be a formative experience: the lessons learned were subsequently used with a vengeance elsewhere.

But the damage this did to Celtel's standing as an operator in Uganda was still reverberating three years later. Interviewed in 2003, the new commercial director of Celtel Uganda admitted publicly for the first time that being the monopoly operator had damaged the company: "When Celtel started we were the monopoly mobile operator and if you look at monopoly operators globally, they all operate in the same way. So, Celtel Uganda did nothing different that was not done, for example, in Finland; I come from Finland. Telecom Finland was the monopoly operator in the beginning and they also extracted higher prices. It is a global phenomenon.

"However, people have long memories here; that is what I have realised (smiles), and one of the critical factors in the mobile industry is price and affordability. We are talking about a market which is extremely price sensitive and that's why a critical aspect of the product offering is to price it correctly and that's why we adjusted our prices to match the market realities. We've lowered our rates and so the price perception, which the market still has, of Celtel being extremely expensive has no reality base."

Further investment by all operators in Uganda was hit hard by the government taxing them at every turn. In 2001, the government introduced a 10% excise duty tax on airtime. This led to a 20% decline in growth of airtime for Celtel. UTL, which by now had its own mobile operation, was also saying that the tax would hit its ability to implement its plans and that airtime had gone down "amazingly." For once all three operators were united against government policy.

For Celtel, it was a crunch issue and at the time it withheld a significant part of its promised $10 million additional network investment. Mo Ibrahim was asked by *New Vision*'s business correspondent why the company's charges were so high. He replied a little testily but with complete honesty: "People forget what the problem is. Every Shs100 we charge a customer, Shs30 goes to the government, 17% VAT and 10% exercise duty and so on.

"Basically we operate in an industry that is expensive by nature. We have never sent a dollar home since we came here. People should be realistic. This is not charity. Business is not built on charity. I need to retain some reward for my shareholders. I think competition is lovely and important because that is the way to force operators to improve their services, lower tariffs and enhance the service. This is the only way to ensure excellent services by competition."

But the government has never reduced taxes on mobile users and Uganda remains one of the African countries where mobile operators are most highly taxed.

CHAPTER 6

Doing business in a war zone

Whilst MSI was acting as junior partner with Vodafone in Uganda, it was off chasing licences across the continent, including in Malawi and Zambia. But its most audacious move was to get licences in three countries that were about to emerge from civil war: Congo-Brazzaville (Congo B), the Democratic Republic of the Congo (DRC) and Sierra Leone.

In 1999 Congo B was still embroiled in a civil war that raged fitfully across large parts of the country. The war had emptied the capital Brazzaville of most of its inhabitants, with the exception of fighters from the different sides and a few brave souls operating the last remaining available services. One of the tallest buildings in the city – the Elf Tower – had, like many buildings, all of its windows blown out. From the correct angle, you could see straight through the building. It seemed as if almost every building was in ruins and everything that could be looted had been looted.

The relative calm of the afternoons were frequently shattered by artillery duels between the warring armies. However, these seemed to take place between 3pm and 5pm as if responding to some invisible schedule. From the capital of DRC, Kinshasa, just a short distance across the Congo River, it was possible to get a grandstand view of the action.

It was like watching some obscure Hollywood action movie: the Ukrainian mercenaries would come up the river and shoot at the other side, before retreating down the river. But sometimes the war had tragic consequences for the spectators in Kinshasa: poorly aimed artillery shells would overshoot their targets and explode in Kinshasa, killing people who were themselves just beginning to experience the end of their own civil war.

At night in Brazzaville, the humid air was tortured by the sound of soldiers often high on drink and psychotropic drugs firing off call signals with their machine guns. One would fire two short bursts followed by a long one and in response another soldier would fire three staccato bursts. Occasionally, the racket from these call signs would last for hours.

The city was swathed with informal checkpoints set up by the combatants, including children. This made travel in the day hazardous but extremely dangerous at night. Children with guns are an odd combination of childish whims and perverse adult authority.

The war destroyed much of the country's infrastructure. There was no operational road between Brazzaville and the country's second city Pointe Noire. The only way to travel between the two cities was flying in the aged ex-Soviet military Antonov 26s which were extremely accident prone. The entire country's communications backbone was destroyed and very few telephone exchanges were operational.

Congo B had been created by a French land-grab in the mid-19th century and eventually became a colony known as Moyen (middle) Congo to distinguish it from its neighbour, the Belgian Congo. At independence in 1960 it named itself after the Italian born French explorer Pierre Savorgnan de Brazzaville who founded the capital city in 1880.

Celtel employed Mamadou Kolade, a Senegalese citizen, to journey to Congo-Brazzaville to acquire an operating licence. When Kolade first arrived, the president had given his daughter a mobile licence but she had no money to do anything with it. As Mamadou Kolade explained: "I went to the president with the difficult task of explaining to him that his daughter had neither the expertise nor the resources to make use of the licence. Yet he was refusing a licence to Celtel who could bring support from the World Bank (through its investment from the IFC)." Without communications, and due to its poor security status, the UN was refusing to let its personnel enter the country.

Eventually the president was convinced by Mamadou Kolade's arguments and Celtel paid a $0.5 million licence fee as a sign of its intent. However, the day the agreement was signed there was an extraordinarily large afternoon exchange of artillery fire. But Kolade believed things would improve: "Conditions were very difficult. But I believed that the country would get through its civil war and solve it in the end. I believed in the country." After Celtel invested, the security status of the country was upgraded largely because it was now easier to communicate with.

Getting the licence was one thing. Actually launching a network and operating it was another. In late 1999 Celtel appointed Rob Gelderloos as the managing director for the country with the responsibility for making things happen. Mamadou Kolade picked him up from the airport in one of the city's few cabs and took him to the only operating hotel. For $250 a night, there was a bed with no mattress, there were no windows, no running water, no lifts and no restaurants. But as this was really a sellers' market for the only hotel, to complain would be futile. Rob Gelderloos had spent many years in Africa but even he was shocked: "There was no infrastructure of any sort. It was the worst situation I had ever seen in my 15 years in Africa. There were bullet holes all over the hotel building."

At an interview for the job, Rob Gelderloos was told that the target was to have 1500 subscribers by the end of the first quarter and 5000 by the end of year five. Gelderloos responded by asking for a higher target: "I thought that this was a ridiculous plan so I said to him let's aim for 20000 in the first year."

Celtel was going through one of its periodic cash squeezes so when Rob Gelderloos arrived the main obstacle was money: "The company had little or no money at that time. Mamadou took my credit card and settled a three-month hotel bill. The project manager took my personal cash, the $5000 I'd bought with me, and started distributing this to the contractors to get them working again. They had stopped working because of non-payment."

But with or without money, the company was under pressure to get something going. As part of its licence agreement, it had three months to launch the network. But buried deep in the small print was a clause saying that a service had to be launched in six weeks or a forfeit would be paid. The project manager had only been in place two weeks. So Gelderloos asked: "How soon before we can launch anything like a network?"

By mid-October 1999 it had erected a single cell site in a fortnight in the centre of town and launched its operation. It started without a billing system and handed out 20 phones to the government to show that it could live by the rules of the licence. Because the day it launched was politically charged in terms of national politics (it was the anniversary of a recent rebellion), it held a low-key ceremony in which the Minister of Telecommunications handed out the phones to the president and his ministers.

Demand for mobile phones was immediate and dramatic. Celtel began with a single shop in Brazzaville as it was the only one that had been finished. On the day it offered network service, about 500 people turned up, pushing and shoving, trying to get in. In the end, the police had to be called to control the crowd. As Rob Gelderloos remembered it: "We sold 750 subscriptions that day. We continued to sell at that rate for the first week and then we ran out of telephones. It was only after that incident that the distributorships decided it was a serious market."

The following two months saw another two base stations installed but there was still no billing system. Growing pains hit the company hard. Celtel Malawi would not sign off on its billing system so the vendor would not ship another one to Congo B until the situation was resolved. Without a billing system, the network operated an innovative but rather frightening business strategy: charge a high access fee but give the minutes away without telling the customers!

It was selling customers phones with access to the network and telling them they had a $100 airtime limit when in fact there was actually no limit. As its only other competitor was charging $1000 for a phone, there was actually plenty of margin. Initially customers would call back and ask: "how much air time have I got left?" The staff were told to say that they would be called back. But it only took users a couple of weeks for customers to figure out that calls were free and by December 1999 the network was so full, no more subscribers could be added.

However, this was not the complete disaster it might have been: the access price charged to new users was sufficiently expensive to allow the company to phase in demand, and its only other competitor had just 3500 subscribers.

Finding money was a continuing difficulty so Rob Gelderloos hit on a novel short-term fundraising scheme: "I had received from head office 200 old-style mobile phones that were both ugly and comparatively heavy. Every time I put a sign up outside the office reading 'Free telephones: $250 each' people bought them. They loved to buy a free phone for $250 and this was the way I was able to pay my staff. I had 10 or so staff at that time." The capital needed for the network had been calculated on basis of having 2500 subscribers but with 15000 subscribers it needed a great deal more capacity. The company had become a victim of its own success and was not always able to fill this gap.

Customers were told that there were shortages of phones and because expectations were low, they understood and were happy to come back later. Sometimes the company had to delay selling the phones because the network was congested – it took 12 months to break out of this congestion/demand spiral. But by the end of the first year it was the market leader with 15000 subscribers, a great deal more than its original estimate. When the billing system was finally in place in February 2000, it was actually able to charge US25 cents a minute but as Gelderloos saw it: "Even if we had charged US50 cents a minute it would have been chock-a-block." It was also able to implement a pre-paid card system, building on the lessons learnt in Uganda.

The company opened up coverage in Pointe Noire but as project director Tomas Jonell found out, even apparently easy things were far from simple. He decided that they would have a grand public occasion when the new equipment arrived in the city: "We'd rented three big Iluyshins to ship in three base stations and a switching container. In order to get permission to land, you had to pay in cash for the gasoline (for your return flight). So we sat there with the press, the minister, and the TV cameras and watched as the plane circled, turned around and flew away. I had to tell everybody to come back at a later date. Someone had failed to pay for the gasoline.

"When it finally did arrive, they opened up the front of the plane and said where can we offload? Haven't you arranged transport? We found a truck with a wooden base but there were no cranes. So we had to pull the container out of the plane by hand, dump the containers on the ground and then roll them on tree trunks. What normally takes two to three hours, took about two days."

And although a third operator (Orascom) entered the market in June 2000, Celtel had established its position as market leader and was able to hold on to it. At this point, there was no interconnection agreement between operators so it was not possible to call people on other networks. Rob Gelderloos kick-started the process: "I sat together with the manager of the Orascom operation. We agreed on a free interconnect between the three operators, and this was the reason for our growth."

Two and half years later the civil war was over. Brazzaville had been partly rebuilt and the road to Pointe Noire had been reopened. The country was preparing for elections and the population of the city returned from the bush. By this stage Celtel had 125000 customers which would rise to more than one million customers.

Mamadou Kolade was next involved in Sierra Leone, another struggling African country emerging from civil war. Celtel had won the licence in a competitive bid during a brief lull in the fighting in 1998. However, as Terry Rhodes recalls: "We had to enforce the force majeure clause in the licence. This is in every contract and licence but is never used: but this time was different as the country was again embroiled in civil war. We left our equipment and took the people out. Amazingly enough when we went back, the equipment was still there."

After a long succession of coups, Foday Sankoh's Revolutionary United Front had set off a civil war that saw warlords fighting over the country's plentiful supply of diamonds. After a brief period of elected government, a military free-for-all ensued with atrocities committed by all sides. Child soldiers would ask their victims whether they wanted their arm cut off at the wrist or the shoulder. Initially a West African peacekeeping force led by the Nigerian army reinstated the elected President Ahmad Tejan Kabbah, and in mid-1999 a cease-fire was signed making Foday Sankoh vice president.

Whereas Congo B had people with skills who just happened to be fighting, Sierra Leone seemed to have lost its entire skilled workforce, who had fled overseas. But as with Congo B, there was no infrastructure worth the name. The only working communications for civilians were satellite phones.

In early 2000, when Kolade landed at Freetown's Lungi airport, he found the Nigerian peacekeepers were out in force. The airport lies on the other side of a large bay and when he arrived the transport options were limited. The bay no longer had a bridge across it. There were a fleet of ex-Soviet M18 helicopters that it was rumoured had been stolen after the collapse of the Soviet Union. Initially there were five but over time three of them crashed. These are still in use and another of them crashed in 2007.

The other option was to take the ferry or a hovercraft. Sometimes the ferry's engine would break down in mid-stream and the boat would drift out into the ocean. The hovercraft was another kind of experience altogether. The roof to the passenger area was made of clear Plexiglass through which the sun beat down mercilessly and there was no air conditioning. Also it didn't so much hover as bump up and down on the water.

In order to operate, everything had to be paid for in cash. Mamadou Kolade found himself carrying $10000–15000 in cash on his trips: "Something like that can go in your jacket pocket but it's very dangerous. When we paid for the licence, we had to bring in $50000–100000 and you had to rely on people who could hide it. There was no Western Union or banking."

Hotel accommodation was again limited and was without windows or electricity. Kolade stayed at the Mamba Point Hotel which was the base for the shady traders who exchanged goods for diamonds. He met one of these people, an Indian woman whose business was to buy rice for the different factions in exchange for diamonds: "One day she showed me a handful of diamonds and said: 'how long will you have to work for Celtel to get this amount of money?'"

There were others who had also obtained mobile licences but none had started operating. Celtel's plan was to launch in Freetown as it was, relatively speaking, one of the safer areas in the country.

Negotiations soon followed which produced a ceasefire but the part of Freetown that Celtel was based in was occupied by the soldiers of Foday Sankoh's RUF. The soldiers thought that mobile phones would be enormously useful for military communications and threatened staff to hand them over. Mamadou Kolade remembers that they had to find a hundred polite ways of saying no: "Some of his guys used to come to Celtel to demand free phones and we had to find diplomatic ways of saying no." But this strong desire by the soldiers to have mobile phones meant that they didn't destroy the network equipment that had already been installed.

Celtel Sierra Leone's first managing director was a no-nonsense, ex-Royal Air Force officer Howard Martindale. He took a dim view of the constant interruptions and petty bribes required to get through the militia road blocks across the city. But as they were simply wooden poles slung across the road, he'd accelerate his 4×4 vehicle and drive straight through them. One day some of the militia put up a metal and concrete barrier and he wrote off his vehicle trying to drive through it.

Celtel had a national licence and it planned to provide national coverage. But a minister told Celtel that it couldn't enter one particular area where they wanted to extend coverage. The minister said: "It's controlled by people who are not in the same party as us." Luckily Celtel was later able to use the help of Vice-President Solomon Berewa to sort it out.

Under the 1999 agreement the West African forces were to be replaced by UN forces, but when the Nigerians pulled out in April 2000, Foday Sankoh's RUF started violating the cease-fire. In early May, just as the network was due to be switched on, there was a demonstration against the violations outside Sankoh's house in Freetown, which was only a few hundred yards from the Celtel compound. The Celtel and Ericsson engineers spent the day huddled together on the floor of the living room slowly getting through a crate of whisky whilst outside the RUF opened fire, killing a number of local people.

The British government intervened, sending in troops from a warship moored in the harbour and at one stage the Celtel engineers were flown out by helicopter to the warship.

Celtel struck a deal with the government whereby the UN forces and the government took over the operation of the cellular network for the length of the dispute and paid a monthly, all inclusive fee.

Sankoh and other senior members of the RUF were arrested and the group was stripped of its positions in government. Sankoh later died in prison awaiting trial for war crimes.

On reclaiming its network in September, Celtel quickly added 2000 subscribers but, as in Congo-Brazzaville, the new arrangements over interconnection were to prove problematic. The incumbent telco Sierratel had almost no infrastructure and few paying customers. But it found itself very quickly owing Celtel a very large sum of money. After many months of not paying, a deal for repayment was negotiated, which although not always honoured, was, according to Mamadou Kolade, better than no payment at all: "Eventually we got them to recognise the debt and agree to pay us $10000 a month. They would pay some months and not others."

There was no telecoms legislation and no independent regulator until 2006 so the government had a free hand to make things up as it went along, changing the rules overnight when it needed money. It imposed a 10% excise duty on all calls and made the change retroactive. When Celtel protested, there was no room for manoeuvre: "It said you should have made sure the subscribers paid it. But we said how can we pay what we haven't recovered? If we have to, we'll need to raise the price of calling by 10%. And, of course, the government says no you can't do that."

To this day the Sierra Leone government is still capable of maverick behaviour. Having liberalised the telecoms sector and encouraged Celtel and others to build international gateways to improve connectivity with the outside world, it decided in 2006 that there should only be one (state owned) gateway and the mobile companies had to mothball their satellite dishes.

Mamadou Kolade's next port of call in the hunt for licences was DRC. This sprawling former Belgian colony is larger than Europe but has almost no tarmac roads. After the fall of its last dictator, Mobutu Sese Seko, this vast country was fought over by competing militias and armies from several neighbouring countries. In 1997 the father of DRC's current president, Laurent Kabila, marched into the capital Kinshasa with assistance from Rwanda and Uganda. Kolade, who had been in Kinshasa before this change of government, now had to make himself known to a whole new set of ministers: "They had literally just arrived and they didn't know about telecoms. The Mobutu people who had made up the government just fled the country. The new government took over their houses and cars."

Because of the civil war, the local currency – the Zaire – was almost worthless. A taxi ride into the city from the airport cost 1.5 million Zaire, almost a briefcase full of notes. Every transaction of any scale in local currency was like an election count. The choice of accommodation was better than in Congo B or Sierra Leone but came with its own dangers: "You had to rent a house but make sure it didn't belong to some former government person, who might at some stage return and ask for it back."

The new government was being feted by a number of people who wanted to do business in this vast country but few were prepared to commit themselves financially: "We went to see the Kabila people and the person in charge said, I am fed up with people who come to have their picture taken with me and leave and never come back. If you're serious, you have to make a down payment of $2 million."

A few days before meeting President Laurent Kabila, the Minister of the Interior sent his people to detain Celtel's management: "They said you're in breach of the law because you've been currency speculating. They were detained in Kim Mazieres prison. From there they were taken to the office of the Minister of the Interior. There were complaints about smuggling dollars and speculating. We were released a few hours later. But when we met the president we had to say to him that it sent a bad signal to investors to arrest people like this. It was a time when anybody could arrest somebody else for no good reason and it was difficult because there was no chain of command when you tried to get people released."

Mamadou Kolade argued that the government had already given a licence to a Gambian businessman for $2 million (who set up an operation that later became Vodacom) so why not give them the same treatment as a reputable international company? But because of the level of interest in mobile licences, the government realised it could hold out for more. Eventually a figure of $15 million was agreed with a member of the government, payable over the period of the licence. But the president himself disagreed and wanted the sum to be paid over four instead of 15 years. This was the deal that was signed.

Kolade felt he was in hot water until this point: "Celtel HQ was angry because it had already paid US$2 million and had not yet got the licence. I spent whole days in the ministry waiting for the minister or his advisers. I was at the president's office so often many people thought I was a presidential adviser."

Deals in this kind of situation do not go through conventional channels: "Often I would use the chief of security to get messages across to the president. I would say to him: 'You need to show investors coming to the country that their investment will be secure.' One day he said to me, 'let's meet at 7am.' He was dealing with security issues so he was 10 hours late for the meeting. But when he came, he said it was a done deal. We paid US$2 million for the licence fee. This same guy later went to jail and was condemned to death because they think he killed the president."

President Laurent Kabila senior was shot by one of his guards on 16 January 2001 – his death was announced a couple of days later. Mamadou Kolade was summoned to DRC for what were described as security reasons: "We had to come to DRC for security reasons. We came in a chartered jet to speak to the security people to make sure our investment and our people were secure."

The president's security people had worked out that those who had plotted the president's assassination had used Celtel's mobile phones. "At this point, the security people said to us, the people who killed the president used Celtel phones. We want the recordings. We had to tell them that the network does not make recordings but could identify who had called whom and at what time. They wanted to make sure that Celtel was not involved in the coup. At that time a lot of things in DRC were irrational."

But the real problems, once Celtel started to roll out beyond the capital, were the militias and the lack of roads. When, a couple of years later, it implemented a network in the east of the country, it had to deal directly with the rebel leaders to make sure it could be done with their permission. Everything had to be flown there and this could be fatal: "We had a plane full of equipment that crashed. It couldn't land in Kisangani so it decided to go on to Rwanda and land there. When it landed, its tyres burst and the plane exploded and all our equipment was lost in the fire."

One of the "up-sides" of the deal with the government was a series of tax and duty exemptions. This meant it could get to market with much cheaper phones. The existing players, Starcel (previously part of Telecel) and two others, had old analogue networks and were charging $1200 for each of its phones: "We said we were going to sell phones for less than $100. We said this to the minister and he didn't believe it. But after the tax and duty exemption was signed, we made sure that he was one of the first people to get a phone."

These existing networks had around five base stations each and operated in different parts of the capital city, Kinshasa, so those customers who could afford phones were largely forced to carry one from each network. Against this background Celtel launched with 25 base stations and coverage across Kinshasa, and by the end of 2000, approximately one month after launch, had 18000 customers. One month later it was larger than the other three operators combined.

In order to be able to operate in DRC Celtel had to bring money collected to a central location because the banking system was not working. So the scene was not unlike a casino counting its cash. There was a large room about 20 metres by 20 metres that was full of cash laid out on tables, being counted by rows of women, wearing face masks to protect them against dust from the notes. "We had to take money from the central office to the bank. It put an enormous risk on those carrying company cash. Luckily they didn't get attacked as people never discovered we were moving cash."

In time, all three countries recovered from their civil wars. Back in Sierra Leone, Celtel had connected the diamond town of Kono. Before its arrival in July 2004 there had only been one rather unreliable and expensive satellite phone and people were forced to travel to Freetown to receive calls. Almost immediately, there were 3000 customers in Kono alone.

Mama Fatmata, a fish-seller in the beach-front village of Tombo, told *Panos Features* in the same year: "I can now not only do business with ease and speed, but also call my relatives abroad whenever I want to." A mobile phone meant she could deal directly with her customers – mostly Lebanese retailers – without going through middlemen.

As security returned to the country coverage spread to an increasingly large number of communities. In 2006 the company paid $5.3 million in taxes to the government, in a combination of licence fees, excise duties and income tax collected from its employees.

In DRC, as with Sierra Leone, coverage expanded quickly right across the country as the security situation improved. However, Vodacom entered the market in December 2001 by buying into an existing operator, Congolese Wireless Networks.

At the time Vodacom was the largest operator in South Africa and suddenly woke up to the fact that its competitor in South Africa, MTN, was building a substantial business elsewhere on the continent. Vodacom had launched in Tanzania in 2000, and was, at the time, doing well without substantial local competition.

Vodacom came to DRC with a strong brand and strong financing but initially struggled to make its mark, curiously using English marketing slogans in a French-speaking country. However, it radically revised its tariffs in mid-2003, and over a period of three months overtook Celtel to become market leader – although Celtel was later to grab back this position through what was perceived as better branding and more effective marketing.

Surviving in these kinds of civil war situations requires a very special set of qualities and a particular way of doing things. Reflecting on this, Mamadou Kolade thought: "It was important to have good intelligence about what was happening. In Congo B, key contacts would tell me: don't go out today or don't go to this or that place. We developed a system whereby we would ring the French Embassy and they would tell us the safe areas to go to."

Celtel staff had to know how to handle themselves when dealing with child militias. Kolade continued: "In Brazzaville, the militia men were frequently on drugs. So you're dealing with a 15-year-old who would fire a rocket propelled grenade up in the air for fun. What goes up must come down and when it does it can kill people. At the road blocks, these armed children were very frightening. You have to know the country very well. Which tribes are against which other tribes? I did my military service in the south of Senegal when there was a separatist rebellion going on so I had some experience of these things. But in these kinds of situations, you need a lot of good luck."

But wasn't all this enormously dangerous? "Celtel did say very clearly to me, don't take risks. I would not do it again. I did it because of Mo and his vision of a pan-African company. It's hard to believe but it's true. I wanted to show that we Africans could achieve it."

Although Mo Ibrahim was persuasive, he was also careful: "The level of civil strife was not as intense as later. Things were not that bad. I wouldn't put people at risk. We could have won a franchise in Iraq but I wouldn't go there."

Mo Ibrahim was also making the point about horses for courses. He and his team understood the risks in Africa and could operate successfully there. As we shall see later, MTC Kuwait, Celtel's current owner, does indeed operate successfully in Iraq.

CHAPTER 7

Steering clear of corruption

Africa has been one of the most difficult places on the planet to do business in. It has not just been about its smouldering civil wars and lack of infrastructure but also a toxic cocktail of corruption and political risk.

Whilst developed nations are not immune to corruption, its existence on the continent is, as the UK government's 2005 Commission for Africa report noted: "... systemic in much of Africa today ... It is another of Africa's vicious circles: corruption has a corrosive effect on efforts to improve governance, yet improved governance is essential to reduce the scope of corruption in the first place. All this harms the poorest people in particular."

Sub-Saharan African countries top the Transparency International Index of corruption and although no accurate figures are available, in 2002 the African Union estimated that it cost the continent $148 billion or one-quarter of the continent's collective GDP.

A US court case brought against the now defunct Titan Wireless usefully illustrates the pattern of this kind of bribery. In 2005 US military and intelligence contractor Titan was fined $28 million for paying a $2 million bribe to the 2001 re-election campaign of President Mathieu Kerekou of Benin to secure the purchase of the country's incumbent telco OPT with a dowry to carry out capital projects. The bribe was to secure a higher price for the deal. There is no suggestion that the president of the West African state was himself aware of any wrongdoing.

The deal was part of a disastrous foray by Titan Wireless into the African market at the height of the telecoms boom. It formed a joint venture with Benin's fixed operator, OPT. Under the terms of what was described as a Build, Operate, Co-operate and Transfer project, it was to install and operate a GSM cellular network, a rural telephony network, a fibre optic backbone and local telephone switching equipment.

Titan admitted making illegal payments through a former employee of Titan Wireless. Titan said in court that in February 1999 its president and chief executive officer, Gene W. Ray, had met with the president of Benin, Mathieu Kerekou, to announce plans for "a state-of-the-art communications system" for the country. Gene Ray had been Titan's chief executive since the company was founded in 1981.

The action bought against the company stated that from 1999 to 2001, $3.5 million flowed from Titan "to its agent in Benin, who was known at the time to Titan to be the president of Benin's business adviser." About $2 million went to the president's election campaign, the commission said, some of it to buy T-shirts bearing the president's picture. President Kerekou was re-elected in March 2001 with 84.1% of the vote in a race against one of his own ministers.

So when Celtel decided to build its own pan-African mobile network, it was faced with a choice – to bribe or not to bribe – and decided to "do the right thing." This had been decided from the outset in the company's values and confirmed in 1999 when the company had deliberately decided to accept the consequences of pursuing World Bank and development bank financing. The "pirates" might follow the traditional African route of spreading money about to get things done. Celtel would very publicly take a different road.

One particular incident in a West African country where it had purchased a licence from another company clearly set the company on this path. Initially Celtel bought this licence from another company that had failed to get an operation up and running. It soon became clear why. There were a number of outstanding issues that had not been formally sorted out, including the licence fees, an interconnection agreement and a waiver on import duties.

A decision was subsequently made that the then CEO Sir Alan Rudge would make a confidence-boosting visit and sort out the outstanding issues with those concerned. A letter was sent to the ministry outlining the issues that the company needed formally resolving.

Sir Alan Rudge was met by the new local agent whose first question was: "Did you get my fax before you left?" This, it transpired, was in essence a list of people who needed to be paid off (with payments of up to $50000) simply to set up meetings with them. Rudge ignored this "local custom" and simply arranged a meeting to discuss things with the ministry.

He was assured that everything could be sorted out but found himself with three senior civil servants at the end of the meeting all looking expectantly at him. After a lot of small talk, they made it clear that if he made the payments agreed, they would "sort things out." He said that he wanted everything in writing. The most senior of the civil servants remarked that this would not be possible but they would have an understanding. If there were any problems, he should bring them directly to this senior civil servant. It seemed clear that having already gutted one lucrative fish, they were attracted by the prospect of Celtel becoming their next victim.

On his return, Sir Alan Rudge confronted the board and recommended that the investment be "put on hold." The company wrote off the $750000 it had already invested in the licence and local office. This country was among the last to attract mobile operators and there were similar kinds of disputes with other operators who tried to invest. Far from "oiling the wheels of business" as its defenders maintain, corruption in Africa is often just about taking money and not actually getting anything done.

Celtel's board was consciously put together as a group of people who understood the issues and could act as a firewall against pressures to work in a less transparent way. As Mo Ibrahim explained: "Some might regard such a heavyweight board as restrictive to a start-up company. But for Celtel this has helped navigate some of the complex political currents. We made it clear that any requests for political donations and the like would be referred to the main board and discussed by the representatives of major donor nations. It showed everybody that we were serious about our anti-corruption stance and it was a great protection."

The whole process of acquiring a mobile licence starts with finding a local partner and if you're not willing to bribe then this requires a great deal of finesse. As former CEO Marten Pieters explained: "If you're looking for a licence and you're trying to select a local partner, you have to be very careful. There were occasions when we broke off contacts with some potential local partners because the influence they might have had would be bought at too high a price. We did not pay 'commissions.'"

It is always best to get successful local businessmen who can also put some money into the partnership. We always made sure that the local partners were not too closely linked to politics.

In some countries, there were several political factions that made up the government: being seen with one would not win influence with the others. It was a fine line to tread but the spread of varying degrees of democratic accountability across the continent meant that simply finding the right presidential crony was not, in most instances, going to be enough.

The local partner needed genuinely to be a part of the local community rather than simply the "front man" for a particular set of interests. The company would be very vulnerable if it was represented by a "front man." As Marten Pieters saw it: "You're vulnerable because of the amount of money you bring in. As a business, you're there to stay and so you have to obey the rules. So you need local partners who can open doors but who are also independent."

Finding these kinds of local partner was not easy, as buying a 10–15% shareholding in a local Celtel operation might cost between $10–15 million and very few local businessmen had that kind of money. Nevertheless, Marten Pieters felt that Celtel had been lucky in its choice of this kind of local partner.

If you didn't arrange things by finding a local partner as shareholder, then it was possible to employ a consultant to mobilise support and advocate for a licence bid. The problem with this approach is that on occasions people took the money and ran. In another African country, the company paid a respected local businessman to play this role and he simply did nothing. Whichever way you approached it, things could get murky.

In yet another West African country, there was a local partner who claimed to have a licence but it turned out he didn't have what he claimed. He was neither paid nor given shares so he went to court and the matter was eventually settled in Celtel's favour.

If the process contained all these elements then Celtel's decision not to get involved with bribery might seem either foolish or brave. In fact it was neither as it was simply a combination of principles and pragmatism. For in the first instance, none of its investors would want it to be involved in this kind of activity. As Marten Pieters explained: "We were clear with our shareholders, which included blue-chip institutional investors like ING and public sector investors like the World Bank's IFC, that you don't want to be in the news as part of a corruption scandal. Therefore we will never entertain these discussions. If we go to a government, we explained to the shareholders, we will do it in a professional way."

According to Terry Rhodes, this clear stance actually strengthened rather than weakened the company's negotiating position: "The country where we handed back our licence was very useful to us because thereafter you could point to a concrete example. You could have had a successful business and we had all the elements lined up but we didn't do it because there was too much corruption. Taking this stand gave us enormous credibility. We used the example without naming it in other countries. We would tell governments that if we had that licence we couldn't pursue for these reasons and clearly they did not want to be in the same position."

As Terry Rhodes explains, it was important for the company to be seen to set a high standard and be in it for the long run. "We started by saying that we're going to run this company as you would a First World telecoms company. Then after the WorldCom case in America, we used to say that we were going to run it better than American companies. Let us do business clearly and transparently. Let us import stuff fairly. Let us take money out to pay our shareholders. We'll pay taxes fairly and we're here to stay."

Much is said against the World Bank, but Rhodes identified its involvement as important in countering corruption for the company. "The World Bank link was important as a back-stop. After the Cold War, it's the World Bank that was trying to modernise these economies. Nobody likes it but they have a lot of clout. Big institutions also carry a lot of weight. Would they intervene on your behalf? No, that might upset the apple cart and they wouldn't want to get involved in a commercial dispute." So whilst it might carry weight, if push came to shove it still meant that the company had to be very smart in its dealings.

Nevertheless there has been a strengthening of corporate governance in the area of bribery that has provided a firmer line of defence. The USA passed the Foreign Corrupt Practices Act outlawing this kind of bribery in 1977 and as the Titan case illustrates, it has been known to take action. Other countries were far slower to react to this problem. For as Mo Ibrahim often points out when dealing with the subject, until 1997 bribery was still tax deductible in some Western countries.

That said the OECD countries implemented the Anti-Bribery Convention in 1997, with entry into force on 15 February 1999. But implementation has been slow and there have been few prosecutions. The UN Convention against corruption was adopted in October 2003 and came into force on 14 December 2005. A total of 140 countries signed but only 52 have ratified it though not all G8 countries have participated.

Mo Ibrahim draws an analogy with adultery: it always takes two parties. It also takes two for bribery to be successful: the corrupter and the corrupted. So this stronger stance is now also supported in parts of Africa. The former president of Nigeria Obasanjo created an anti-corruption commission and it has pursued a number of high profile investigations. Also under the auspices of the New Partnership for African Development (NEPAD), African countries are undertaking peer reviews, a key part of which is to reflect on issues like corruption and governance.

Often avoiding corruption involves little more than not allowing you to be sucked in. As one country manager experienced it: "Sometimes you were told jokingly 'my bill will come at election time' to see how you would react to it. If you didn't blink or react, they didn't expect anything. It was a standard shakedown. All they could do was withhold a signature but they couldn't do that forever."

But Celtel has not only sought to occupy the high moral ground, it also endeavoured to change the terms of engagement. When some kind of bribe was asked for, those working for the company tried to turn things around. If the chief of a village wanted money to allow a base station to be erected, why not offer to spend money improving a local school? Making a contribution to a local community meant you knew where the money was going and that it would reach local people.

The distinction sounds like a fine line but the difference was clear as one of those involved remembers: "When we went into eastern DRC, we were perfectly happy to appease the rebels in that area by contributing to reconstruction of that area as part of our corporate social responsibility programme. But it was not personal corruption."

Because mobile phones were extremely popular with Africans, Celtel could deliver a different but equally valuable contribution to politicians rather than a bribe. "What would sometimes happen is that the president would say to us I'm up for election and I need $50000 for the presidential re-election fund. We would say we can only do that if it is approved by the board. Of course it would never get that far. We want to help so why don't you open say a cell site for us and you'll get all the glory bringing communications to the people who lived in the new coverage areas."

Corruption is far from simple and those that study it in detail talk about "need and greed" corruption. Oiling the bank accounts of national politicians fits firmly into the latter category: for example, the late Nigerian dictator Sani Abacha reportedly sent more than $1.3 billion out of his country to London banks. Those in the "greed" category are generally selling access to decision-making or the making of a particular decision.

The "need" category of corruption is everyday folk on low salaries asking for bribes to supplement their income. It's the policeman who lets you off an "on-the-spot" fine. It's the customs officer who lets your bags go through unexamined. It's the lowly civil servant who ensures that your application is dealt with quickly. The phone engineer who gets your phone installed ahead of where you are in the queue. A Nigerian phrase describes it rather well: "dash me"; in other words, grease my palm.

One area that particularly affects those doing business in Africa is the importing of goods. Customs officers are able to delay goods almost endlessly unless paid off. But even these could be avoided with some determination as one country manager found out. "We used to say 'hold your horses' to them as these sort of delays were included in the lead time. I also used to get local staff to go and talk to the customs officials. Find somebody from each tribe to go over and say 'my job depends on it.'"

But this was the hardest area to make absolute decisions about, as Marten Pieters acknowledged: "The greyest areas are in small corruption. You have to deal with the reality on the ground. If the police stop you in certain countries, you might pay a $2–3 fine just to get moving. That's a grey area. Another example would be getting construction permits. But we have fired people who have stepped over the line."

So where was that line over which employees could not step? Everyone from the top down knew that the company wanted "to achieve sustainable development of telecommunications businesses in Africa in an efficient and profitable, responsible and ethical way." It stated that its key value was integrity and that it believed that it was possible "to conduct a business across sub-Saharan Africa that follows best business practice." Indeed it stated that it would itself adopt "best practice for corporate governance and expect the same from governments and partners." The last word should perhaps go to the IFC arm of the World Bank, who awarded Celtel its first ever "Client Leadership Award" in October 2004: Celtel is "a company that sets the gold standard for its peers anywhere in the world, a company that is a role model for others, regardless of sector, region or country" (Peter Woicke, IFC Executive VP, October 2004).

CHAPTER 8

Tanzania's TTCL – the deal that took five years to complete

In mid-2000 buying Tanzania's incumbent TTCL looked very tempting to some in MSI. The company came with a dowry, the new mobile licence that would get it into the fast-expanding Tanzanian market. The attraction of the deal was always clear from the start. It was the last opportunity to get into Tanzania which was shaping up to be an extremely interesting market. At the time there was no other route into the mobile market in Tanzania – there were already four operators and one of its South African rivals Vodacom was already there doing very well. MSI had on its staff a senior manager, Omari Issa, who had worked on the TTCL privatisation process for the World Bank. By the standards of African telco incumbents, TTCL was considered to be a good business.

But the warning signs were there from the beginning. For the first and last time in its life, MSI was buying and would have to manage a fixed line operator, something that it had never done before. The deal took the company to the brink of a complete breakdown in relations with the Tanzanian government and it was five years before a final agreement was negotiated.

Initially the main obstacle was that some members of the board had difficulty imagining themselves running a fixed line operator of this kind. As Terry Rhodes saw it: "There was a lot of agonising. Some of the board said you're mad. You're not setting up from scratch and suddenly you're responsible for 2500 civil servants."

"We were horrified by the due diligence on the fixed side. The few computers that were to be found in the business were just gathering dust. But it appeared to have assets and a business." Whatever the downsides of the business, the temptation was its scale: this was much bigger than previous MSI deals and was seen as a major step on the long trek to becoming a convincing pan-African operator.

In order to prepare itself for this leap into the dark, it teamed up with Deutsche Telecom's consultancy arm, Detecon, who, for a large management fee, would run the fixed line business. Thus, freed of that day-to-day responsibility of running TTCL, MSI then saw it being able to do what it did best: running a mobile operation. Furthermore by the time the TTCL deal was completed, MSI had a CEO, Sir Alan Rudge, who had been number two at British Telecom and was thus familiar with the management of fixed line operations. But with hindsight, Terry Rhodes conceded that there was always only one objective in his mind: "I always wanted the mobile side of the business but needed to go via the fixed side of the business because it was the only way to the target."

The Tanzanian government launched the tender to privatise TTCL in June 1999. A 35% share of the company was put up for sale to a strategic investor (who would have management control) and letters of invitation went out to 120 organisations. Nine companies submitted pre-qualification documentation but only six of them were pre-qualified. These were: MSI and Detecon; an Indian consortium consisting of Mahanagar Telephone Nigam Limited (MTNL), Telecommunications Consultants India Limited (TCIL) and VSNL; Mauritius Telecom; MTN; Sasktel (who in 2007 took over the management of TTCL on a contract); and Vodacom with WorldTel. In the end, the Indian consortium and MTN chose not to submit bids.

MSI and Detecon bid the highest, offering US$120 million, which was to be invested in the company, and 800100 connections by 31 December 2003. The nearest bid was almost half that amount: Vodacom and its partner bid $66 million, whilst Sasktel offered $60.5 million.

Over the next few years MSI was criticised regularly for the price it had bid but the rationale was simple: as Rick Beveridge noted: "We bid $120 million for 35% of the company which is a valuation of the full company of $343 million – but the money didn't go to the government, it went to the company, so the $343 million is 'post money' – before we put the money in we were bidding a value of $223 million 'pre-money.'

"For that $223 million 'pre-money' valuation MSI got a bundle of assets: notably a mobile licence which was the same as the one for which Vodacom had, a few months earlier, paid $50 million and, the government assured us, a bundle of more than $80 million of receivables. If we assumed that only $60 million of that $80 million would ever become cash we still bought the fixed operation for around $110 million, which was one year's revenue or four times the 2000 expected EBITDA. It was, on paper, a good deal."

However, MSI's bid was seriously higher than the next bid and for a moment the team worried that they had overpaid, until they had a courtesy visit from one of the losing bidders a month or so later. "They asked us how we had arrived at our bid and it soon became evident that our valuations of the basic business were similar, and they had been working on the deal and done more technical due diligence than we had, but then we got to explaining how the valuation had translated to a bid price, the difference between pre-money and post-money valuations and they looked like they were about to be very ill, they just hadn't got the financial experience to think in those terms."

On privatisation, TTCL with its strategic partner – the Detecon/MSI consortium – was to be issued five new licences, one for each of its different services: fixed, cellular, data, radio paging and Internet. In order to safeguard the fixed line business, TTCL was granted a four-year monopoly on fixed line telephony. But more importantly for Celtel, it joined the other four mobile operators in the country: Mobitel, Tritel, Zantel (at that stage only operating in Zanzibar) and Vodacom Tanzania.

The deal was welcomed warmly in Tanzania by the executive chairman of the Parastatal Sector Reform Commission: "The Detecon/MSI consortium is well capable of helping TTCL play a significant role in the development of the telecommunications sector in Tanzania. It has valuable experience, excellent skills and adequate resources which through its control of the TTCL board and management will enable TTCL to strengthen its financial performance and improve its customer relations. The ultimate winner is the consumer of telecom services."

But the warmth was soon to disappear. Once the MSI acquisition team got its collective head "under the hood" a number of rather scary unknowns emerged. The company had not completed its audit for the year 2000. The number of outstanding bills, particularly from the government, was very high and it was unclear whether this revenue could ever be collected as some of the bills went back four years.

There were other problems that were not made public. As a result of tied aid, TTCL was buying all its network equipment from a mix of different suppliers and at list price rather than with the customary discount a telco might expect. The collection of money from outstanding bills was exacerbated by a billing system that seemed almost designed to create confusion. The incumbent telco's customers were used to challenging their bills. However, if you challenged one bill and agreed to pay a second one, the billing system simply put the money for the second bill against the disputed first bill. So if you were querying a bill, there were all sorts of muddles that might delay payment.

On the government side, officials assured MSI that the bills would be paid and that the company was about to have a much better year financially than it had previously experienced. With these rather Panglossian assurances, the government was pressing for immediate payment of the full $120 million bid price.

MSI explained that the original deal had been written on a business model that reflected what they had been shown. In its turn, the government of Tanzania assured them that there was $80 million worth of "receivables": in other words, bills that would be paid. MSI could not really say "we don't believe you" as this would be a sign of bad faith.

So MSI chose to create a purchase strategy based upon how much of the actual outstanding money got paid. Acquisition team member Rick Beveridge summarised the essence of the deal offered to the Tanzanian government: "Since the accounts are not complete, we are prepared to believe you but if we cannot collect (the money), the company will be worth less on the basis of a formula agreed. And we needed the accounts independently audited – not just the draft accounts used for the purposes of the privatisation proposal."

Huddled in a corridor with the government bankers Rothschild, the MSI acquisition team offered this as a way out of the payment impasse. As Terry Rhodes recalled: "We said how about we split the payment into two? We'll pay $60 million now and a further payment based on actual results. If the accounts receivable are less, then we will pay less based on a formula that was a multiple of EBITDA based on the verified accounts. We agreed this right at the last minute, little expecting that it would take three years (to reach an agreement)." MSI/Detecon paid the first tranche, a sum of $60 million, in February 2001 and promised to make the second payment once the 2000 accounts for TTCL had been agreed. Meanwhile MSI's shares in TTCL were held in escrow as security for the final payment.

MSI's in-house lawyer on the team, Gretchen Helmer (now Jonell), found herself working on this negotiation for much longer than she imagined: "Tanzania kept me busy for five years. We agreed that the second tranche payment would be calculated based on an EBIDTA multiple on the audited accounts. We foresaw potential differences of opinion with the Tanzanian government in the calculations, so I insisted that any dispute be settled in an international forum and that English law play a role.

Our concerns were realised when the Tanzanians delayed the audit of the accounts but were asking us to pay the (full) $120 million."

It also set up the cause for the dispute that was to rumble on for so many years. The Tanzanian government had in its mind that if the revenue from the outstanding bills was largely collected, it would still be seeing a second payment of $60 million. The privatisation process in Tanzania had been the subject of much political debate and if the government came away with much less than this sum, then it would suffer a serious loss of face. In its heart of hearts, it probably believed that MSI could pay anyway and really the financial problems were for it to sort out. At the highest level, Tanzania's government could not accept that the outstanding bills could not be collected. MSI suggested that if the government believed this to be the case, why didn't the government set up a team to collect the money? The proposal was not accepted.

But if the main dispute with the government was over the size of the second payment, there was also the debate over MSI's service obligations that also played to the wider political gallery. Its original service obligation was extremely carefully worded. As Rick Beveridge outlined it: "We would expand the network to a total of 800100 direct access lines. There was an allocation that so many were fixed and so many were mobile and the rest were by any technology that worked best. If there were three lines into a building and these served 100 extensions, it was to be counted as 100 lines. It was carefully worded and the vast proportion of lines would be provided by any technology, mobile or fixed."

But when the country's regulator, the Communications Commission, translated the sale agreement into a licence, it became an obligation that rested solely on the fixed line business.

With these raging political arguments going on around it, MSI realised it had bought a company but had little real control over it. Also whereas a team of 20 external managers might be needed for an acquisition of this scale, MSI was just getting by with a much smaller team. The number of staff working for TTCL far exceeded the number required for its rather modest number of fixed lines. The MSI/Detecon management tried to continue the redundancy programme started by the government but it got mired in court action and the political maelstrom swirling round MSI's ownership. Worse still, a large section of the relatively modest number of fixed lines needed to be disconnected because of long-standing unpaid bills.

MSI insisted that Celtel Tanzania was a separate company with separate management (although wholly owned by TTCL), but this meant that there was no sense of shared vision between the two operations. Indeed the majority of the TTCL staff seemed to believe that Celtel was just another competitor taking customers that were rightfully "theirs," whereas Celtel, being asked to use TTCL infrastructure wherever possible, thought that TTCL were abusing their position as preferred supplier.

For example, on one occasion in 2003 the transmission people at TTCL phoned their Celtel opposite number to say that a critical transmission link had to be taken down for a couple of hours. It was arranged that the work would be done at first light on a Sunday morning to minimise disruption to Celtel's network. However, TTCL decided not to pay for accommodation close to the site, which meant that the engineers only left home mid morning and arrived on site just in time to take down the transmission for the two peak hours of the day – with the result that almost all of Celtel's revenue for the day was lost.

TTCL had not really got to grips with computing either, despite being a communications company of some scale. Kamiel Koot recalled a particular incident when he went to see them during the signing of the deal: "I went to see the acting director of finance. He had a dustcover on his PC. I said what's that used for? Accounting entries, he said. How do you get information? We send people to collect it from the regions otherwise people wouldn't send it to us. Why not run an automated system? We can't get the data lines. At this stage, the company was supposed to have a turnover of $110 million."

All of the contentious issues between the government and MSI became the subject of megaphone negotiation through the media initiated by the Tanzanian government, and there were plenty – including competitors – who were willing to stoke the fires of disagreement. For Tanzania's politicians, it was much easier to skirt round the uncomfortable truth about the unpaid bills and to paint MSI as another unreliable foreign company that promised much but delivered little. In an attempt to "up the ante" in these negotiations, the government called MSI "gangsters" and threatened to take back MSI's shares in TTCL.

The government appeared to have MSI well and truly over a barrel. Its shares in TTCL were held in escrow pending the second payment and MSI seemed to be in a very weak position. The government appeared to be winning the propaganda battle as things got messy and heated between the two parties.

MSI and its London lawyers argued that the agreed jurisdiction in the event of disagreement would be the UK; the London courts agreed and MSI was able to obtain an order from a London judge that if the Tanzanian government took its shares, then the UK court would authorise the seizing of the Tanzanian government's assets in England. The case set a historic precedent: the court allowed MSI to keep its injunction secret and only serve it if (or rather when) the Tanzanian government started to take MSI's shares in TTCL. In addition, the company's lawyer, Gretchen Helmer, was sworn to an oath of silence as the judge did not want the injunction to influence the negotiations.

With the Tanzanian government threatening to seize MSI's shares, the decision was taken to serve the injunction. But there was every chance the paperwork for the injunction would be seized as Gretchen Helmer entered the country. So an elaborate smuggling route was set up and a rather unlikely courier was employed to carry the documents in: "I had to meet the guy who was to carry the documents at a bar in Nairobi. I was really mad because the guy got cold feet. Eventually he was persuaded." Luckily for Gretchen Helmer, and the company, the arrangement worked because on arrival at the airport she was searched for well over an hour, but the secret cargo made it through.

However, at the negotiations when the injunctions would have been served, no one turned up: there had been either a leak or surveillance had revealed what was about to happen. Gretchen Helmer went off with two of the larger members of the negotiating team to serve the injunctions. She was scared but found herself taking advantage of the fact that she was a woman. "We got into the relevant government building at 4.30pm and everyone was still working. We pushed our way past two armed guards. The idea was that, as a woman, the guards wouldn't touch me so I managed to get through. In this way we served the injunctions on the Minister of Finance, the Minister of Telecommunications and the PSRC. It said, 'if you touch our assets, the courts will imprison your nationals and impound your property in England.'"

Arrangements were made to devise several options of leaving the country as there was some chance that the government might detain or arrest the negotiating team members. Out of this brinkmanship came the agreement to set up a Joint Working Party to resolve the outstanding issues.

In mid-2002, a rather buffeted MSI issued a statement to the press in an attempt to set the record straight: "The difficulties … for both MSI and the Tanzanian government have revolved around establishing a set of fair and reasonable accounts to measure the financial performance of TTCL in the year 2000. When the new management at TTCL began to review the accounting systems and data, a number of serious issues were revealed. In particular a very large proportion of the company's recorded revenue in the year 2000 and before was never collected, and the age of much of the debt was such that it was never likely to be.

"The total uncollected debt due from non-government sources amounted to 75 billion shillings ($94.8 million) at the end of 1999, rising to 115 billion shillings ($144.1 million) at the end of 2000. In addition a number of other issues emerged relating to interconnect payments and ret-roactive changes in taxation rules which also impacted upon the financial performance of TTCL in the year 2000.

"In February 2002, to avoid further debate on the validity of these issues, MSI proposed that a third party independent review of the accounts, or a joint investigation team, would be desirable. Following lengthy con-sultations, the government chose the latter and a joint MSI/Government Taskforce was put in place on 9 April 2002. The Taskforce confirmed the issues, but in the time available were unable to agree their resolution. On 31 May 2002 it was agreed by MSI and the government of Tanzania that a further month should be granted to resolve the accounting issues and that a joint project team with an independent chairman is formed to carry out the work. Until this appointment is made MSI have agreed that the govern-ment should provide an acting chairman."

MSI's CEO Sir Alan Rudge described the appointment of the chair of the group: "The Joint Working Party found all the errors in the accounts but the chair had not been appointed. I didn't know any of the suggested names so I accepted an accountant. In my mind, accounts were accounts and an accountant would take that attitude. But as soon as he was appointed chair, he started doing the accounts all over again and came up with a set of accounts showing that TTCL was sufficiently profitable to justify the remaining $60 million payment.

"These accounts were then presented to me and I asked: 'I'm supposed to accept these accounts?' So we were in a stalemate again and this went on for a couple of years."

MSI enlisted the help of the British High Commissioner and the World Bank to explain to the Tanzanian President that if the matter wasn't sorted out, it was unlikely that another foreign investor would be willing to put money into the country. The president asked his chief of cabinet to get the matter sorted.

Sir Alan Rudge was at the first meeting to tackle the issue. "So we had a meeting around a big table surrounded by civil servants and Tanzanian and London lawyers. Anything we said was drowned out by the government's lawyers. There appeared to be no way out. So I said to the chief of cabinet would you mind stepping out of the room to have a conversation with me. I said to him, 'I'm an honest guy. I do everything openly and I honestly want to resolve this one.' He told me what the lawyers had already cost him and how they were making millions keeping another dispute going. So we set up arbitration. We agreed the process and an independent expert. It took six months and every time there was a blockage I went to see him. Those six months were spent drafting the expert's terms of reference."

The government was adamant that the set of accounts produced by their Joint Working Party chairman would be the basis for the expert determination. MSI remained convinced that his work was fundamentally flawed. Gretchen Helmer observed that "these negotiations became heated and often personal because the stakes were so high. At one point the Tanzanian government sought its own injunction to prevent the accounts being finalised, but MSI would not be bullied."

The process agreed was that each side would prepare its case for expert determination by a senior London partner at KPMG who would decide on the merits of the cases submitted. The expert found in MSI's favour but there was still a surprise, as Sir Alan Rudge discovered: "We thought this would mean we had nothing to pay but the decision still meant we had to pay just over $5 million (including interest). The figure was important because if we'd been forced to pay the (full) $60 million the other side was expecting then we'd have gone bankrupt." A joint announcement accepting the ruling was made at State House the next day and MSI paid the following day. The whole process had started in 2001 and was not concluded until the end of 2004.

In one of life's unintentional ironies, MSI was named as best Dutch foreign investor at the end of 2003. The award was presented by the SMO Foundation for Business and Society and was judged by a panel of experts including the Dutch government, development agencies and the private sector.

But even though the accounts had finally been settled and the second payment made, MSI (now Celtel) were keen to get out of running TTCL after its bruising encounter with the Tanzanian government had made it all but impossible for it to run the company.

Essentially TTCL's network investment had been long delayed by the legal case, and whilst everyone pledged publicly that they would work together to make TTCL a success, by this stage Celtel had had enough.

So the Celtel team moved to the next set of negotiations: unbundling the mobile operation from TTCL. In November 2004 Celtel signalled to the government that it wanted out and asked the government to propose a share price so that it could buy out the government's interest in the mobile company. So finally in 2005, it was agreed that Celtel would own 60% of the new mobile company, Celtel Tanzania, and retain a 35% share in TTCL but simply become a sleeping partner.

After the lengthy negotiations in Tanzania, Celtel never seriously looked at buying a fixed line operator again. As one of the team involved put it: "It's still a point of pain for us. We saved ourselves money but we got lots of bad will." It may seem strange but if you buy a "national asset," it takes you right into the arena of national politics.

Arguably it lost the political war of words but since Celtel today has over two million mobile customers in Tanzania, it certainly won the commercial battle.

CHAPTER 9

Snatching Kenya from under the nose of rivals

Celtel's next major deal was to have a more successful outcome and was to signal to both competitors and investors that it had moved up a league. It was no longer the plucky outsider but one of the three main challengers on the continent along with MTN and Vodacom. When Kenya's mobile operator Kencell went up for sale in late 2002, it was widely assumed that the aggressive South African operator MTN would be the favourite to make a quick acquisition because they had narrowly lost out on the original licence. Celtel had already tried three times to get into the Kenyan market, but without success. But as is sometimes the case, things did not turn out right for the favourite.

Sixty per cent of Kencell was owned by Vivendi who had been involved in an abortive attempt to take over Celtel before the crash in telecoms and Internet companies. Under the leadership of the flamboyant Jean-Marie Messier, Vivendi had become a conglomerate that seemed to be snapping up everything in sight from Hollywood studios to telecoms companies. It had gone from a boring French utilities company to one of the dazzling stocks of the telecoms and Internet bubble.

Sceptics might carp that they couldn't see what it all added up to but the company seemed to grow and grow. However, when the market turned against Vivendi, it had to sell off many of the companies it had accumulated in its earlier shopping spree, to pay off its debts. It had expanded into Africa as part of a global strategy during which it had acquired the majority stake in Moroccan incumbent Maroc Telecom as well as Kenya's second mobile operator Kencell: it paid $55 million to acquire the operating licence. However, from 2001, its local partner Naushad Merali saw them as "absentee landlords." In May 2002 Vivendi took a decision to get out of Africa.

In September 2002, one of its executives phoned Naushad Merali to say that it was placing its 60% shareholding up for sale. Naushad Merali, the CEO of a Kenyan conglomerate whose interests stretched from tea and coffee plantations to tyres and energy generation, controlled the remaining 40%. But he also had pre-emption rights over the shares Vivendi was selling: in other words, he would get first refusal on buying them if he could match or outbid any buyer Vivendi found. This made him a key player in the sale from Celtel's point of view.

The company that was the target of Celtel's interest was number two in a market with only two players. It had 1.2 million customers against

Safaricom's 1.8 million. But even though it was second, its number of customers and the potential size of the market would make it a considerable catch. If MTN were to prevail and capture a significant positioning in Kenya, it would have the key markets in South, West and East Africa and its position at the top rung of the continent's wireless industry would be virtually unassailable, relegating Celtel to permanent second tier status. Tsega Gebreyes, the former Celtel board non-executive director who had assumed executive responsibility for business development in April 2003, observed that Celtel could match any deal offered to Vivendi by MTN.

This rather remarkable assertion was noticeably at odds with the industry's appraisal of Celtel's capabilities and, not surprisingly, the company's dark horse campaign got off to a sluggish start. From beginning to end, the strategic successes and reversals in this campaign would largely hinge on the question of credibility – could Celtel really pull this off, could it succeed in its bid as management insisted?

In the second quarter of 2003 when Celtel made its initial overtures, Vivendi was not inclined to believe Celtel could finance an acquisition of such magnitude. Their judgement, which in this matter was widely shared, could hardly be faulted. Curiously it was a factor which would work in Celtel's favour. The deal was worth around $250 million and MTN, the putative buyers, also doubted that Celtel had deep enough pockets to play at this level, allowing Celtel a more open playing field than it might otherwise enjoy. Celtel management and board members would readily acknowledge the formidable odds against them, but those odds, whilst compelling, were not daunting. Celtel took the bold and very costly step of engaging Lehman Brothers to sort out the financing requirements.

Without any clear idea as to how the financing would ultimately be arranged, Celtel pressed on with little success for months. Despite appeals at executive and board levels, Vivendi rebuffed invitations to engage in a serious negotiation, but the Celtel resolve would not waver. The persistence would finally begin to pay off in the summer of 2003, when ironically it was a former Vivendi executive who introduced Mo Ibrahim to Naushad Merali. Naushad's significance to the process had always figured prominently in Celtel's strategic plans. Anyone hoping to travel the road to success in the sale negotiations would eventually have to obtain Nashaud's blessing. Celtel knew that Naushaud could be a critical ally, especially given Vivendi's focus on MTN. As Dave Hagedorn, who worked in the deal, explained: "The key was Merali. He had pre-emption rights that enabled him to have the first option on buying the shares if they were to be sold so if we could (get him to work with us and) fund these pre-emption rights, the company was ours."

In other words, Naushad's preemption rights represented a final opportunity to outflank the competition in the Kencell bidding process.

At Mo's meeting with Naushad another important ingredient apart from the strategic benefits was added to the mix. Mo immediately liked Naushad; he appreciated what Naushad had accomplished and how he had done it. Naushad was similarly impressed by Mo. The two agreed in principle to work together. Very shortly thereafter, their two teams set up a meeting to hammer out a deal through which it was eventually determined that Nashaud's group would have 60% of the new business and Celtel the remaining 40%.

Their plans, however, were of scant significance to Viviendi. In early 2004, having semi-publicly hawked its shares around potential buyers, a Vivendi executive phoned Naushad and told him the company was going to sign an exclusive contract to sell its Kencell shares to MTN. Having secured this understanding with Vivendi, MTN was less than circumspect in the way it approached Naushad. Indeed, those with long memories could not help recalling the difficulties South African Breweries had with its acquisition of local Kenya brewer Castle.

MTN's attitude irked Naushad: "They came to see me (for due diligence purposes) and I said I've got pre-emption rights but their attitude was, we've bought it and it's ours. They were speaking down to us." He was later quoted in the press as saying that if MTN had sent its chairman Cyril Ramaphosa, the deal would have been done. But in March 2004 a Vivendi executive flew to Nairobi to confirm personally that an exclusive contract had been signed, selling its 60% share to MTN for $230 million. Naushad Merali pondered this and eventually told Vivendi that he was reserving his position.

This was the opening Celtel desperately needed. Having made up his mind, Naushad would have 20 days to exercise his rights and then 20 days to make payment. Celtel would not have more than a 40-day window to arrange financing and close the deal. And the deal had become more expensive. The sale price had escalated above initial expectations and the share percentages had been reversed: Celtel would now get 60% and Naushad Merali 40%.

Complicating Celtel's position was its knowledge that MTN was also now talking to Naushad. Whilst Celtel and Naushad had outlined a deal, they had not signed a legally binding contract. Had Naushad signed such a document and Celtel failed to meet its financial obligations, Naushad would have been a business partner with MTN, the group he would have openly spurned. Celtel did not provide a break fee to Naushad which would have given him greater certainty of Celtel's ability to perform. In the absence of such an agreement or legal documents both sides were taking a risk on the other's performance.

Mo remained convinced that the requisite trust had been established and a deal between them would be done. Naushad Merali is still surprised that anyone should have expected him to sign anything: "If there's trust, there's no need. The beauty of Mo (Ibrahim) is that he's a superb character. He's very friendly and when he decides on something, he goes for it. He told me that I would be able to join the board of Celtel International and he was as good as his word."

Others on the Celtel deal team were mildly reassured when Kencell's CEO Philippe Vandenbroeck warned the press in January 2004 that the deal with MTN was far from complete: "Nothing has been done. Any noise coming from anybody is just rumours."

Having tucked away both the financing and Naushad's verbal agreement, Celtel's acquisition team turned its attention to the clock. As Dave Hagedorn recalled: "Our advisers Lehman Brothers were incredibly focused and well organised. As there was only 40 days to complete the deal, Lehman had all the days from one to 40 printed up on A5 and we would find ourselves coming in and turning the number over. It made you very conscious of how little time there was and it was a great motivational tool. There were lots of details to be sorted out in 40 days."

Many hours were spent haggling over the wording of the letter of guarantee that would go to Naushad Merali. But eventually the documents for the pre-emption payments were sent off to a Vivendi office at the very last moment.

According to the interview he gave to *Balancing Act* (issue 210) Naushad Merali said: "I signed the sale and purchase contract with Vivendi at the board meeting at 7.00pm on Monday night. The place was packed with lawyers." The price was Sh18.4 billion ($230 million), calculated at a dollar price of Sh80. He now owned 100% of the company. At the same time the Celtel acquisition team was waiting patiently in a nearby room. "At roughly 9.00pm I sold the shares to Celtel for Sh20 billion ($250 million), although there were sizeable fees to be paid to all the advisers."

Tsega recognized that the main factor sealing the deal was not really money. Naushad Merali had already made his own fortune. He was attracted by the idea of an international company that had Africans in its top management. He was attracted by the idea of the Celtel International board and joining a community of his peers. He had also developed a relationship with Mo Ibrahim and the two understood each other. This was a textbook example of soft issues driving results beyond hard numbers, a key factor in Celtel's success when it eventually tackled the Nigerian acquisition and IPO trade sale.

The Kenya deal was an enormous boost for Celtel. CEO Marten Pieters believed it made the company more credible in the eyes of investors: "At that point we were operating in some of the most difficult countries in the world and these were places that didn't ring any bells with investors. Kenya was different because people had a picture of it through things like tourism and safaris. It just lifted our whole portfolio." More doors would now be opened in the financial world when the company went looking for banking relationships. Vendors and prospective partners would view Celtel in a different, more serious light, as it had now become a major player on the continent.

Changing the shape
of the problem

There's an underlying paradox in the development of Africa's telecoms markets. Everything that happens elsewhere will happen in Africa but at the same time Africa is different from everywhere else. Just like everyone else, Africans wanted a mobile phone service and often a fancy phone to go with it. But because this was Africa, they used the service in different ways.

One example is the widespread practice of "beeping" between mobile phone users in sub-Saharan Africa. Beeping involves calling a number and hanging up before the mobile's owner can pick up the call. The phone's user gets a missed call message with the caller's number. This can either be a request to call the user back immediately or a way of signalling something like "pick me up now" or something more abstract like "I'm thinking of you." Africans have taken the technology and adapted it to the realities of the income they earn.

Those who arrive on the continent thinking it will be almost the same as Europe or North America but "just a bit different" find it a graveyard for imported bright ideas. This kind of loose thinking very quickly bumps up against a number of stark realities.

To start with, Africa's mobile markets are dominated by pre-pay customers: well over 98% of the millions of African phone subscribers pay in advance for their service and often in small denominations. At the other end of the scale, there are the post-paid subscribers who are very small in number and are often corporate users. Things are changing slowly but when Celtel started there was no pre-pay and no African middle class to speak of in most countries. All too often those who could left to go overseas to make their fortune.

Therefore what mobile operators like Celtel needed to do was not create products and services for bored elite users in search of distraction but find ways of targeting those with very little money with services that fitted the patterns of a very different life. In 2002, business academics C.K. Prahalad and Stuart Hart identified these markets as "bottom of the pyramid:" they have estimated that there are four billion people who live on somewhere between $2–4 per day. The majority of people in Africa fall into this category. The challenge for multinational companies is to deliver products that are useful to them at a price they can afford. Celtel was, in effect, implementing a "bottom of the pyramid" strategy even before the term was invented.

The other reality was one that affected African companies wanting to do business across the continent. If you wanted to travel from one side of the continent to the other, it was often easier to fly via London or Paris. In the same way, phone calls and e-mails often went on a long journey internationally to Europe or the USA before coming back to Africa, which was economically efficient, but the distances involved often meant that users experienced delays in hearing the other person and their voices would be heard as echoes in the background.

On a more personal level, entire groups of identical Africans who speak the same language are spread out over the rather arbitrary borders of a former imperial age. For example, Wolof-speaking Gambians (in the tiny strip of land that is the Gambia) are completely surrounded by Wolof-speaking Senegalese. Hausa speakers across West Africa travel frequently between countries, as either seasonal migrants or traders. The same is true for a significant number of Swahili speakers in East Africa. Lesotho is completely surrounded by South Africa and many of its inhabitants travel backwards and forwards to South Africa for work.

The reason that things are structured in this way is partly a function of history. Patterns of communication mirror the imperial past where the main routes for travel and messages were between the capitals of the colonies and those of their imperial masters. There were few routes between African countries, particularly where the imperial power was different.

So in order for travelling business people, traders and seasonal migrants to communicate, they needed (as many still do) to get their phone unlocked, and buy a local SIM card and some local airtime. For those doing this regularly, it often meant that they needed to have two phones to avoid the roaming changes and the fiddly process of constantly exchanging local SIM cards.

But whether rich or poor, most people still carried out a large part of their daily business with cash. Very limited retail banking exists and the number of people using it regularly is always a relatively small minority of a country's population. Very few people have access to credit cards. So whether it's paying an electricity bill or buying an air conditioning unit, it's usually all paid for in cash.

If you're a mobile operator in Africa, you can either say "well that's how it is" and keep taking the lucrative revenues from simple voice and SMS services or you can choose to try to do something in business terms that will begin to change the shape of the problem. From an early stage, Celtel chose to be one of those companies that wanted to find a different way of doing things that would help overcome some of the barriers both individuals and businesses faced on the continent.

New business ideas come in many different ways but often they simply flow from looking at how things work. The idea for Celtel's payments company Celpay, which was launched in 2000, came out of just such a process of observation. Celtel's management discovered that its Malawi operation had run out of pre-pay scratch cards. It decided to dig a little deeper to find out what was going on.

It emerged that not only were low denomination value scratch cards running out but that also, unusually for the African market, the high denomination value scratch cards were also sold out. As it turned out, the scratch cards had been given dollar values and people were buying them as a hedge against the falling value of the local currency, the Kwacha.

Interesting, thought management; so pre-paid scratch cards have become a form of currency. Bank notes are only accepted as tender for buying things because everybody trusts the value printed on them. However, if everyone suddenly believes that something has a value that can be exchanged, a new currency is created.

The second incident that caught management's attention was in DRC. A couple of senior managers went to Kinshasa to meet the father of the current president of DRC, Laurent Kabila, who was subsequently assassinated. For reasons now lost in the mists of time the meeting did not take place. But whilst they were being driven from the airport, the car got stuck in traffic at a junction and they noticed something strange.

There were two guys sitting at the side of the road with a massive stack of currency in front of them, talking into the old-fashioned, "half-brick" mobile phones. Puzzled by the sight, they enquired as to what was going on. It turned out that this was the banking system in a country where the banks had all but ceased to function.

So if you wanted to make a payment in Mbuji-Mayi, for example, the guys on the phones would ring their trusted contacts until they found someone who wanted to make a similar payment in Kinshasa. Once both ends of the transaction were in place, the men by the side of the road at each end would make the payment.

So the Celtel managers decided to arrange a meeting with DRC's Central Bank's governor, who explained that if you wanted to transfer the same money using a bank it might take up to a month if you were lucky but it might not happen at all. Not surprisingly, the Central Bank's governor was enthusiastic when they asked him for a banking licence to transfer money using SMS text messages.

As it turned out, Celtel actually started Celpay in Zambia because it wanted to trial the business in a slightly less demanding English market before launching it in DRC. It wanted to provide what it thought would be an invaluable service for Africa. It won a *Wall Street Journal* award for innovation in financial services.

Setting up the payment system using SMS was incredibly simple. In order to get the payment system, the user needed to purchase a new, larger, SIM memory card with the application on it. This bought up a Celpay menu on their screen which provided easy instructions such as "see account," "pay" and so forth. The user would send a five digit PIN number that was encrypted for security purposes, and then go to the payment platform on the phone and check that the PIN number was correct. An instruction would then be sent to either make a payment or check the balance of the account.

But nothing was quite as simple as it seemed. One of the managers involved recalled: "we started with the retail idea. You would be able to pay your bills with it. But we hadn't thought through the complexity of it. In order to move from cash to an electronic system, there has to be an element of trust. There has to be a change in people's mindsets." At an individual level, this was to prove hard to acquire. A thick pile of dirty notes in your hand or some new-fangled SMS system from your mobile? Unfortunately most people wanted to stick with the devil they knew. If you weren't able to show the money, it was hard to believe that it existed.

At this point, Celtel realised that maybe it was being too ambitious. Because of the trust issue it was a hard marketing sell to the relatively small number of customers who had bank accounts. But it also had to persuade those accepting payments in this way that they would be charged 1–2% of the total payment. Since credit cards and point of sale equipment were not widely known in the country, those accepting payments were somewhat resistant to the idea of being charged to get the money that they previously used to be able to get directly in cash.

So Celtel decided that it would concentrate on corporate clients. Lots of large companies operating in Africa delivered goods to customers and the drivers would go round and collect the money for the goods in cash. As the round of deliveries progressed, the driver became a security risk. Celtel discovered that South African Breweries had been working on the problem in Zambia for two years and were willing to pay 1% for just such a service. Also petrol company BP were making deliveries to petrol stations worth up to $20000 a time and were interested, along with a local cement company.

When it rolled out the service in DRC, it was also selling it as an easier way of making transactions. For if everything's in cash, someone, somewhere, has to sit down and count it. Just buying an expensive meal in some countries takes on the feeling of an election count when you come to pay with your small mountain of notes. Counting like this takes time.

The hapless delivery driver in Kinshasa would perhaps spend over an hour of every day counting the dirty, small, denomination notes he was given. It would then be taken back to a large secure room full of women in dust caps and gloves who would count the money all over again. Celtel was pitching to companies that they could avoid all this time and trouble with an electronic payments system.

One of the successes in DRC was that cash could be transacted against airtime. On this basis, bank accounts in the country doubled from 20000 to 40000 as a result, giving some idea of the potential the system might have in the future.

On the corporate side, the problem was that in the process of distribution, there was corruption. So, for example, if there was a shortage of a particular commodity and the merchant wanted to get a priority delivery, they would bribe the guy who made the delivery allocations. The mobile payment system was so transparent that it would have made corruption of this kind all too clear.

Also Celtel was working with the chief financial officers of multinational banks. Africa was the testing ground for those starting out and as they were frequently moved on to other countries, they were reluctant to make big changes during their tour of duty. For some banks, this was a major decision that would need the approval of head office.

Ironically it also had a great deal of difficulty working with other mobile companies. The larger ones did not want something that came from a competitor and even working with Celtel was akin to what one of those involved described as "brotherly hate." It really needed independence to be able to work with all mobile operators.

In Zambia there was already a debit card and point of sale infrastructure and this caused difficulties on the retail side. Zambian retailers didn't want to have to buy another point of sale device. But there was no similar point of sale infrastructure in DRC and as a result it got 20000 subscribers and transacted $200000 a day.

But probably the most significant use of the system was for making payments to demobbed soldiers for Conader (DRC's National Commission for Disarmament, Demobilisation and Reinsertion) and the World Bank at the end of DRC's civil war. They wanted to pay soldiers as if they were in the regular army for one year to give them time to find alternative employment and make sure they did not go back to war.

The first payments were for $110 to each soldier and were called Orientation payments. The Orientation payment phase of the project lasted six months and took place in Conader Orientation centres that were located near major population centres.

At the same time it was paying so-called reinsertion payments to up to 150000 ex-combatants who were located right across a country larger than Western Europe. These reinsertion payments were carried out primarily by means of Celtel airtime resellers who used the Celpay dealer application for electronic top-ups. The dealer would be sent the money by SMS. He or she would type in the number of the soldier and a PIN and if both were correct, the dealer would pay the money to the soldier. The reinsertion phase lasted for over 18 months. Although some payments were made in other ways, Celpay was responsible for paying out $50 million worth of these payments over the life of the project.

However, financial services take time to build up confidence and Celtel sold the loss making company to the South African bank First Rand in March 2005. The sale was part of the clear-out of non-core businesses before the company's IPO.

A mobile cash payments system remains a tantalising possibility for Africa. All over the continent, users have already cottoned on to the fact that if they can send units they've bought to other users, these can be converted into cash by the person to whom they have been sent. But no one has yet progressed from this crude but effective workaround system to something that would work in a more formal way. At the time of writing, a number of new M-cash services have been launched but it is too early to know whether they have been successful. So Celtel was probably ahead of its time in offering this service.

The next innovation from Celtel had a more lasting impact on the market. The Congo River is just over 3 km wide and separates the two capitals of Brazzaville (of Congo B) and Kinshasa (of DRC); people from both cities travel backwards and forwards by ferry, often on a daily basis. Yet in the early part of the new century there was no direct telephone link between the two countries.

Celtel's local managers in Congo B thought this would be a good idea and relatively easy to achieve. Technically it was going to be extremely simple to put up a microwave link that joined the two capitals: it was possible to see the other capital from either side of the river bank. But it quickly turned into something that ran the risk of becoming an intractable problem. As Celtel's local manager put it: "This was the longest and toughest negotiation in my telecoms experience." But all of the problems came from people and the past history of the telecoms sector.

After the end of the civil war in Congo B, the country's incumbent telephone company Sotelco was in poor shape. It had already lost its international monopoly when Celtel was granted a gateway licence. This threatened the lucrative stream of income that monopoly incumbents like itself had counted on over the years. This income included "international" calls to DRC, a country that was literally on its doorstep.

Therefore rather than start with the incumbent phone company, Celtel approached the minister responsible for telecommunications. It explained that it could connect the two countries and offer affordable rates to users. Not surprisingly the minister liked the idea and encouraged the project. But the trouble started when they approached the incumbent telephone company Sotelco.

Sotelco's managers said no, we don't want you to do it because we've already got a land-based connection. However, this existing connection was actually out of commission. But in line with most African incumbent telcos of the period, it felt that this was their monopoly and therefore that it had the right to be the sole operator. In an argument that became popular with incumbents, it said that if someone else operated the link it would undermine the security of the country. However, in a rapidly changing Africa these arguments were no longer decisive.

Celtel had to find a way out of this negotiating impasse if the link was ever to become a reality. The solution was to offer Sotelco its own link to connect to the incumbent in Kinshasa, whilst Celtel would be given permission to operate its own link as well. With this offer on the table, Sotelco finally gave in and agreed that the link should go ahead.

But the cunning and rather greedy Sotelco management almost upset the whole process by wanting to charge their counterparts in Kinshasa. "We gave both the PTTs at either end of the link the (microwave) dishes. The PTT in Brazzaville wanted to charge its counterpart in Kinshasa for the link. We managed to talk to the PTT in Kinshasa and explain that it had been a gift."

Before, if you called a Celtel customer in Kinshasa from the phone of a Celtel customer in Brazzaville, the call went out on one international carrier's satellite link to a place in the Netherlands, then it went to a central interlink point called Telcoms House in London over fibre, and over another fibre to an earth station place just south of Brussels and then over a different satellite provider's capacity to Kinshasa. It cost Celtel $1 and the user $1.50. With the new direct connection, it cost only US40 cents on the Celtel network, one-quarter of the previous price.

All this was very heart-warming and gained the company many "brownie points" when dealing with governments elsewhere, but did it make commercial sense?

An early sign of the likely impact came whilst testing the link. One of the company's engineers made an error and calls were routed (at zero cost) over satellite. News of this unexpected consumer bonus flashed round Brazzaville by word of mouth at the speed of light. In just two days this free service had doubled the traffic on the Brazzaville–Kinshasa link.

There were two issues with the link. A satellite connection is not of the same quality as a land-based microwave link, due to the inherent delays, or "latency," of the call. The signal actually goes 36000 km up to a satellite and then down, and in this case it did it twice – a round trip of some 150000 km. But there were also capacity constraints on the link. A combination of lower price and increased quality meant that total traffic on the route increased 20-fold in 2003. The manager responsible for putting the link in place estimates that one-quarter of this increase was due to the better quality and the remainder to the lower pricing. Furthermore, as the only mobile network offering low-cost roaming between the two capital cities, this was a major incentive for potential customers to sign up to the Celtel network.

In early 2002 Celtel started to realise that the nature of competition in Africa was changing as the business matured. Whereas in 1999 there were probably 10 or more companies each of which had licences to run mobile networks in one or two countries, by 2002 the effects of the dot.com bust were coming through and some competitors, such as Millicom and Telecel/Orascom, seemed to have stopped all expansion in Africa. The market was beginning to consolidate and four players were emerging ahead of the others in sub-Saharan Africa and becoming a "premiership": these were Vodacom and MTN from South Africa, Orange of France and (they hoped) Celtel.

It seemed likely that over the next three years further consolidation would take place and one or two operators would emerge ahead of the rest of the pack. Celtel looked at this scenario and realised it had a major problem; the dot.com boom years were, by then, obviously over and it was becoming difficult to raise finance, but its three major competitors all had profitable home networks that were throwing off cash, whereas Celtel was barely generating enough cash to keep its major operations growing, and had effectively put its minor operations on the back burner until times got better.

Celtel needed to examine its strengths and find ways to give itself strategic advantages that would enable it to fight off this competition. Some of the clues came from looking at a map of Africa. Compared with the other players Celtel was in a larger number of countries and moreover those countries were largely contiguous – they formed a solid block across the centre of Africa, from the Atlantic to the Indian Ocean, and stretched from the Sahara to Victoria Falls. Further clues came from the experience across the Congo, and the recognition that the borders of Africa were largely lines drawn on maps by white Europeans some hundred years previously.

Celtel decided to turn itself into a "pan-African" company building regional communities and to this end began looking at the obstacles it might face repeating the experience elsewhere on the continent. Terry Rhodes recalls: "We asked the question, if we can have a customer make a call from Kinshasa to Lubumbashi, both within DRC, but linked by satellite, for 25 cents why can't we allow customers to make a call from Freetown (in Sierra Leone) to Blantyre (in Malawi) for the same price."

The answers came down to technology, government nationalism, different taxes and customs duties, regulators and licences, and currency controls, but having set the vision there was a serious effort over the next few years to get the pieces in place to bring the vision to reality.

Having introduced low-cost international calling, the next scheme was even bolder: abolish roaming charges completely in East Africa. The aim was to introduce pre-paid roaming with no incoming call charge for its clients in Kenya, Uganda and Tanzania in an effort to consolidate its position in those markets. As with the Brazzaville–Kinshasa link, the main issues were regulatory and the company had to get ministers in each country to give it their full backing before it could go ahead.

Although the idea had first been mooted in early 2002 Celtel did not at that time have an operation in Kenya and cross-border links between Tanzania and Uganda were over expensive satellite. But in the negotiations to buy Kenya it had recognised that the three countries, which had once shared a common telecommunications company, had the potential to make the pan-African concept a reality. Bringing them together would strengthen each of the Celtel companies, each of which was behind their competitors, Vodafone, Vodacom and MTN in Kenya, Tanzania and Uganda, respectively.

Building the necessary cross-border links to avoid high satellite charges took a few months and getting the necessary licences to be allowed to send traffic over them took a few years, but in September 2006 Celtel launched new tariffs for users of its services in East Africa. The new tariffs were US43 cents a minute during peak hours and US31 cents a minute during off-peak hours to call from anywhere to anywhere in the new roaming region. These rates applied in Kenya, Uganda and Tanzania, and were branded as "One Network."

The decision to have Ericsson as a common core supplier meant the company could use its common call platform across the group and this allowed easier integration of the One Network service. But this technology was used to create a special advantage. As Moez Daya, then CTO of the company, saw it: "When you're visiting a country, you're treated like you're at home. You're part of the family. The product uses the most efficient local tariff and you can top up locally. The concept is very elegant and simple but it took us two years to perfect it."

At the launch, Celtel Kenya's chairman Naushad Merali said that the new tariffs were part of the firm's strategy to reduce the cost of calls on its network across Africa. "The private sector is keen to support the three governments in the East African community, and the new preferential rates for Celtel to Celtel will ensure that businesses are able to communicate cost effectively, efficiently and reliably." In June 2007 it expanded its regional roaming scheme to include three new countries: DRC, Congo B and Gabon, and in November 2007 to Burkina Faso, Chad, Malawi, Niger, Nigeria and Sudan, together an area twice the size of the European Union.

Indeed it was not long before officials from Brussels, struggling to reduce roaming charges for European customers, were asking Celtel how they did it, when the operators in Europe seemed to find it impossible. As the *Economist* magazine put it in September 2006: "Celtel has in effect created a unified market of the kind that regulators can only dream about in Europe."

In a demonstration that Celtel was now winning the political as well as the commercial battle, the president of Kenya said in June 2007: "Lack of adequate communication has been the weakest point in Africa's regional integration. The communication network being provided by Celtel covering East and Central Africa is therefore a major contribution towards improvement of communication in Africa. This is vital for economic, social and cultural cooperation in the wider region."

CHAPTER 11

Building an African brand – the impact of mobiles on Africa

Africa's mobile operators not only changed how people on the continent lived their lives but also how they saw themselves. In a short period of time their extensive branding meant that few people who saw it could ignore it. They began to dominate the "mind-space" of the continent through the aspirational images they projected on the billboards of Africa's towns and cities.

In many ways the challenge for Celtel was that the services offered by mobile operators were strikingly similar. There might be a blizzard of special, short-term marketing offers but prices actually differed little between operators in most markets. Also in price-sensitive African markets simply shaving a little off the price was unlikely to buy long-term customer loyalty. Coverage might also vary between operators but the coverage patterns of established operators were often more or less equivalent.

In order to gain a large market share, Celtel needed to be able to create a brand that people would identify with for reasons that were not purely financial. Branding sometimes seems to be more akin to alchemy than science. Users of products and services form a brand loyalty for a wide range of reasons, both tangible and intangible. Brands as various as Apple, Coca-Cola and McDonald's all have that special magic that makes customers want to come back for more, often despite the occasional failings of the companies concerned.

The template for advertising in the mobile industry was set by Orange's very successful launch campaign in the UK market in 1994 which was subsequently taken global. It combined the use of a very bright orange colour with the slogan that it uses to this day, "the future's bright, the future's Orange." Unusually both its logo and the slogan have survived a change of ownership. It was very different from most technology brand campaigns in that it spoke about what customers might want to use their phones for rather than their technical features.

Orange was originally called Microtel and was the fourth entrant into an already crowded market. Its new CEO Hans Snook hired a British brand consultant to research the best way of addressing the market. This research suggested the company's owner Hutchinson drop the technical-sounding name "Microtel" and instead focus on giving its mobile phone business a more "personal" dimension. Thus, "Orange" – conveying the themes of friendliness, fun, optimism and liveliness that Hans Snook wanted to highlight – was born. So after a great deal of investment and innovations, like per-second-billing and 24-hour customer support, Orange became the UK market leader.

Although Africa had advertising campaigns and branding, particularly in places like Kenya and South Africa, its countries were largely producer rather than service economies. Economic activity was in the main focused on digging either crops or minerals from the ground. The number of people who had the disposable income to buy consumer goods and services was relatively small.

Demand was nearly always higher than supply which meant that advertising and branding were a luxury for all but a small range of goods. There were a few pan-African brands and these tended to be obvious services like airlines and banks. In a world of many shortages, you were lucky to get a service of any kind, let alone be treated as a valued customer. But the more you had to compete for customers, the more you needed a branding strategy, and nowhere was this truer than for the growing number of mobile operators.

But in the early days the lack of branding and marketing skills hardly mattered. For as one Celtel veteran put it: "All you had to do back then was to put up an antenna and people would come and talk." So desperate was the need that researchers looking at rural coverage in one country found that people would climb hills or trees to get even a faint signal at the edge of the network.

So the early years of the Celtel brand were hardly auspicious. For example, when it launched its network in Congo B, it did not consider any billboard advertising. The company that was meant to organise them was simply unable to get anything done locally so the initial campaign was launched in the newspapers and on television. But billboards were to prove crucial in the operator wars that followed. Although literacy levels vary enormously between countries, almost everyone in Africa's main cities would see and understand the images and simple words displayed on the billboards that came to dominate the cityscape.

One of the words on the billboard needed to be the name of the company. This might differ from country to country and in the early days some operators almost had a different brand name for every country they operated in. Local operations would handle what marketing was carried out and the message would be different in almost every country. There was a great deal of tactical activity but no sense of what is known in the jargon as "brand promise."

Ironically early concerns about branding within the company almost saw the disappearance of the distinctive Celtel name. At its first board meeting in Egypt, the company's directors questioned the value and future of the Celtel brand. The board wanted a blue dot rather like the Orange brand but with a different colour. The company did go through an image change but there was never anything very strategic about its implementation.

It was so busy rolling out new operations that things like branding had to wait awhile. Although the "Zamcell" network was relaunched as Celtel Zambia in the autumn of 2001 the network technicians who ran the company knew that as long as they nodded at all the right moments when branding was mentioned they would not be expected do anything very much about it. It was not until Vodacom relaunched its network in DRC with a strong brand image in 2002 that branding became recognised as a serious issue.

The man chosen to take charge of Celtel's branding was Tito Alai. He worked for Kodak in the days of film and paper photos but had cut his marketing teeth working for Unilever. His experience of the communications sector came from working for the owner of Africa's first pan-African ISP, "Africa Online." In between working for African Lakes and Kodak, he had worked as a consultant for what was then MSI Cellular looking at how to position and brand its ISP service.

He was told by the head-hunter who approached him about the Celtel job that the company was looking for someone who was interested in establishing a marketing capability for the company, because as was explained the company had no real marketing function. There was commercial support but this was largely concerned with call centres and customer relationship management. Although it had all the channels in place to respond to its customers, very few of its senior staff had experience of working in what are known in the jargon as "customer-facing" companies. It took about a year to create an appreciation of marketing and the importance of brand building in the long term.

As one insider put it: "This was an organisation steeped in telecoms technology practice," and like most telecoms companies it regarded itself as somehow different from other industries. But as the success of the Orange brand had shown, telecoms were really very similar to other consumer sectors. It was all about connecting customers with services they wanted and creating a brand they felt good about.

Nevertheless Africa was different from Europe. Nearly all of the African subscribers were pre-paid. They paid for their mobile service in cash and the most common purchase amounts were at the small end of spectrum. As noted earlier, this made buying mobile phone time similar to buying cigarettes or small food purchases. In other words, in marketing terms it was in the fast moving consumer goods (FMCG) sector.

To achieve these kinds of sales, Celtel was already selling through two different channels: its own shops and independent resellers who owned their own shops and ran teams of street sellers. It was important to manage the resellers in ways that would help them reflect well on the company. All of these outlets were already branded and to change them to a new brand image was not something that could happen overnight. Any branding was going to be expensive and needed to involve staff at all levels within the company. But once the board accepted that branding would accelerate growth, resources were not a problem.

The first step towards this massive task was not a good indicator of what was to follow. The first agency hired to create the brand were a small agency in South Africa that had no presence in the countries Celtel operated in. It produced initial artwork but there was a clear sense that it lacked immediacy and relevance for African countries. As Tito Alai saw it: "There was a problem of scale and the South African mind-set."

The second agency to work on it was FutureBrand of London. This might seem strange, given that the problem was getting the cultural context right. But, according to Tito Alai, it was a case of adding the African context: "We had a very Afro-centric vision but their chief creative person had never been to Africa. So we needed to do a lot of work to create a very rich brief, pointing out the common threads across the 10 countries (we operated in) so they understood what a pan-African brand needed."

Even with this knowledge, finding common threads across Africa was never going to be easy. It is a continent where 3000 different languages and cultures coexist and thrive side by side. Every country is multicultural but has rigid cultural traditions. Therefore the challenge was how to celebrate this diversity and build this very richness into the brand.

The first exercise at Celtel before giving the brief was to think what it was that people thought was positive about the continent: where could it claim leadership? One of these was in the area of the arts – performance, music and visual arts – where Africa was contributing at a global level. Out of this came the idea that in order to be successful globally, one needed to tell the African success story. This story would be about a successful African business with its customers living fulfilled lives and this idea was built into the brief to FutureBrand. And out of this initial insight came the idea for the strapline: "Making Life Better" (or "La Vie en Mieux" in French) which reflected the company's broader desire to make a difference to the lives of its customers.

One part of creating a brief for branding a company is to look at the kinds of values it would like to reflect back to its customers. Celtel was in a very fortunate position because it was still young enough to be able to achieve what one company insider described as "an extraordinary alignment of values." It got quickly to the point where the company values and the pan-African vision were encapsulated as brand values. For the first time a major consumer brand was created to be both multinational and African – not just an extension of an existing Western brand. The new colours, red and yellow, were chosen to be much more powerful in Africa than the previous blue.

Impact day for the massive brand changeover was 19 January 2004. The parent company name was changed from MSI Cellular Investments to Celtel International. By this time, all of the local operations in Africa except Mobitel in Sudan already bore the Celtel name. The changeover was a huge logistical exercise as the company had 120000 sales outlets. Countries like DRC were larger than Western Europe: there were, for example, over 20000 points of sale in its capital Kinshasa alone. Not only were sales outlets rebranded but all of the company's vehicles had to be repainted in the new livery.

So the branding was carried out in several waves with five countries in the first wave and the balance in a second wave that took place in April 2004. In all, 11 of its 12 operations were rebranded. The company spent $20 million on achieving this change. After it was completed, every company used the distinctive yellow and red colours and the same logo with the strapline: "Making Life Better."

These outward signs were also a signal of greater inward clarity. What everyone working there knew when it was a smaller company became written down as the company's mission: "To be the most successful pan-African telecommunications company." Also the values that had guided the company became codified.

Celtel would be "open, honest and transparent;" see the customer as its most important stakeholder; value teamwork, achievement and success; and be progressive and culturally diverse. But above all: "We do what we promise to do." Whereas this might seem blindingly obvious in the often murky field of African business values, this counted for a great deal. For as CEO Marten Pieters put it in a presentation to a GSM World conference in 2005: "The strength of Celtel will be judged not just by its financial results, but also by the way in which we conduct business – by the aspirations we set and the manner in which we engage everyone whether it be customers, governments, employees or the communities in which we work."

In the same way, the slogan "Making Life Better" was not merely the standard ad agency cliché but had substance, both individually and in terms of the countries the company operated in. For Celtel's phone service provided the means for people to be better connected, to communicate, to stay closer in touch, to share emotions, to get more done and to transact business. But it also meant making an impact across each country, and across Africa, with job creation, skills enhancement and infrastructure improvement.

The campaigns juxtaposed Africa's cultural rootedness with its modern sophistication. Whilst many billboard campaigns still used white faces from European campaigns, Celtel used attractive black people to show successful but believable images of Africans. This was not just about rich urban African sophisticates but real people with real lives. Furthermore one of the company's slogans tapped into the "new Africa:" Exceed Expectations. Athough the campaign was developed centrally; it was translated into local languages for different markets. It also helped develop marketing expertise across the continent. For example, Celtel's marketing support agency, ZX Advertising, expanded from South Africa to each of the Celtel countries.

The problem with this kind of branding exercise is that often the results are hard to quantify – it creates a wonderfully warm feeling for customers but actually generates very little profit on the "bottom line." In Celtel's case, it tracked the incoming results of the branding and found it paid back in terms of both customer satisfaction and market share.

The starkest example was in DRC where Vodacom had come into the market and overturned Celtel's early lead. Vodacom and other competitors had grabbed 47% of the market against Celtel's 43%. But not only was Celtel the loser in terms of market share, its customers were giving it a lower satisfaction rating than its competitors: it trailed 5 percentage points, getting only a 65% rating against the 70% rating for the competitors.

When Celtel turned up the volume with its branding, the competitors may have thought they were throwing money away. But over just one year it regained the title of market leader, increasing its share to 48%, whilst the competitors' market share fell to 42%. With this boost in market share, the customer satisfaction rating went up to 71%, level pegging with its competitors. The branding exercise also had a similar impact in Tanzania.

Perhaps the biggest challenge for the new brand was in Nigeria. As described in Chapter 15, the operation acquired by Celtel was already five years old and had traded first as Econet Wireless, briefly as Vodacom Nigeria and then as V-Mobile. It was rebranded as Celtel in September 2006 and within six months had won the prestigious "This Day" brand of the year.

Based on research commissioned by the company, the Celtel brand was significant in customers' decision to purchase its services. Across all of Celtel's country markets this brand contribution ranged from 25% to 33%. The ability to call (both including network coverage and quality) remained the most important driver of demand.

These two contributions – brand and service – illustrate perfectly that branding alone will not gain a company market share. But all other things being more or less equal (network coverage and quality) it does enable a company to gain a serious competitive advantage.

Celtel gained not only market share and customer satisfaction but also the all-important "mind-share." Celtel, like Hoover and Xerox, was on the way to becoming synonymous with the product. When asked to identify a mobile operator 85% of people surveyed in Gabon, Congo B and DRC named Celtel.

But money spent on branding was also a good investment as the brand itself became one of the financial assets of the company. The estimated value of international brands forms part of the way a company is assessed in financial terms. For example, in 2007 Coca-Cola's brand was valued by Interbrand at $65 billion, Microsoft at $59 billion and Nokia at $33 billion. Not only was the brand value a large number but it was often a significant percentage of the company's overall market capitalisation.

Using the same methodology adopted elsewhere and fitting with accountancy conventions, FutureBrand was asked to assess the value of the newly created Celtel brand. It discovered that Celtel was the leading brand in over 75% of its African markets. It also calculated that the brand value was 2.5 times EBITDA (earnings before interest, taxes, depreciation and amortisation), which gave it a value of over $1 billion in 2006.

Over time mobile operators have come to understand that the power of a brand name was important across the continent as well as in a single country and now almost all operators are moving towards single brand names and brand building of this kind.

Through this process of brand building, Celtel went from being a well-regarded local success story in its different country markets, to a well-established pan-continental brand. It was perhaps one sign that the company was moving from rude youthfulness to a more established position in its field.

But perhaps it is the impact on individuals that is most significant. Emma Sesay, a pregnant woman living in a remote village in Sierra Leone, would have died of delivery complications had it not been for the timely use of her mobile phone. She called her husband who hired a vehicle to take her to hospital in time to deliver a bouncing baby boy. Out of gratitude the relieved couple decided to name their baby boy Celtel, a living tribute to the power of branding.

CHAPTER 12

Raising the funds to fuel growth – cash is king

The hidden story behind Celtel was its struggle to find funding at a time when neither Africa nor telecoms were investors' favourite flavours. Both acquiring new mobile licences and then building networks required an enormous amount of money. For several years Celtel found this cash not by going down the more conventional route of hiring external banking advisers but by hiring its own investment banker.

Once MSI Cellular had been demerged from its parent company, it set about trying to raise money. Whilst the founders injected some equity, it was a classic start-up company, big on ideas and short on money. As Mo Ibrahim himself conceded: "The main problem was cash as there was little money. We needed to start work and there were large amounts of up-front licence fees."

And at first glance, Mo Ibrahim was the first to admit that perhaps his experience with his first company did not really equip him to take on this new launch: "(When I started that company), I had no relevant business experience or even education. For the first three years we operated without a business plan or even a budget. For sure, that's no way to start a business!"

Nevertheless he had learnt the lessons well: "The flip side was that common sense appeared to be the most effective management discipline. We had created a clear leadership free of conflicts and a single command structure with purpose and effectiveness." Slightly wistfully he acknowledged that: "Later we never managed to achieve the efficiency and profitability of the early years (in MSI)."

But Mo Ibrahim was also never entirely comfortable with the bankers needed to raise the money to do things. As one of the original team members recalled: "Shortly after I joined, Millicom approached Mo to buy the MSI software company. This was the first time an external approach had been made. The banks got involved because we needed a valuation. As part of the process we made a pitch to other players. This was all foreign to Mo who hated dealing with bankers."

But beyond the lessons Mo Ibrahim brought with him from MSI, the new company – MSI Cellular – was really started with "chump change." It had small minority shareholdings in Uganda, Hong Kong and India and some money from its founder but little more. With the exception of

Uganda, these shareholdings were partly "payments-in-kind" for parts of the work the original MSI had done as consultancy.

All across the world, mobile licences were beginning to be sold for considerable sums of money; however, the company's initial ambitions were the equivalent of having tiddlywinks to play in the global league. The divestment process from MSI was complicated and took two years but it gave the fledgling company the opportunity to present itself as an entity that could draw on operating experience. It was the increasing attraction of actually doing rather than advising that had motivated Mo Ibrahim to set up the company. Having a privileged position as consultants to the "first-movers," this was a team that believed it could do it better. Whilst some came from traditional incumbent operators, they could see how things might be done very differently.

As Terry Rhodes, one of the founders, recalled: "We were operators at heart. You couldn't help but feel that you could do better. We had done a full global trawl and seen that big sums were being paid for licences in Europe and the USA but also in places like Indonesia. But there were parts of the world that were both too small and too far away to be interesting to investors. Apart from South Africa, hardly anyone had done anything of any scale in Africa." So although the company later was to become a highly focused pan-African operator, this was the moment when Africa presented itself as the most promising opportunity among many others. Over several years, MSI perfected the formula that was to be the making of Celtel: "We started in small countries and war zones. It was our league before we started to move up."

Mo Ibrahim gathered a small team around him to chase licences and find money to pay for them. He had already recruited Terry Rhodes, a former colleague from BT Cellnet, into MSI in 1995. Terry Rhodes had been doing regulatory and strategy work for Cable & Wireless and had seconded one of his staff to work on a new Cable & Wireless project: setting up MTN in South Africa. By this stage in 1995, MSI Cellular had already started its Ugandan operation. Rhodes had been recruited on the promise of shares and even had to put some of his own money in: "We had all been recruited with the offer of shares and we had to put our own money into the business. This was an important principle."

The man they chose to act as "banker" was Kamiel Koot: "I had seen Mo about a deal. He said we really need a chief financial officer. I wanted to be an entrepreneur but Mo thought he wouldn't be able to afford to match my banker's salary. But he offered a package (including shares) and I said yes." Others were more sceptical: someone who had worked with Mo Ibrahim previously told Kamiel Koot: "I don't know why you want to go and work for that tinpot company."

At this stage, the company was an idea in motion but actually had very little to show. As Kamiel Koot put it: "We were licence hunters. Nothing was really operational." When CDC, the British government's development finance institution (later to become Actis), invested in September 1998, it was sufficiently worried about this lack of experience to insist that its former parent company MSI provide operational support.

But any form of success was a long way away as the company scrabbled to finance its first deals. At the point of the demerger, there was $11 million available. A performance bond for its Hong Kong shareholding accounted for $3.5 million, leaving just $7.5 million to play with. Almost immediately, the company faced its first cash crunch.

Through Vodafone, its partner in Uganda, MSI became a founder member of the Misrfone consortium bidding for the licence in Egypt. But what was a major company like Vodafone doing working with a minnow like MSI? According to Terry Rhodes: "It was partly our technical expertise but also that we could span the different cultures. We were almost a mediator in any boardroom. Within the team we could be equally at home with bankers from America or in a smoky café in Cairo."

Originally a 15% partner in the consortium, MSI had been squeezed down to 5% once Vodafone decided to merge with a competing consortium led by the US company Airtouch. (The following year Vodafone was to conclude a $65 billion merger with Airtouch.) At first MSI's role had been seriously scrutinised by the new partners, but they were sufficiently won over to let MSI lead the bid team for the marketing and technical submission.

As Terry Rhodes recalls: "We were trying to coordinate across different time zones and different cultures, through Ramadan and Christmas." And in a sign of things to come, there were some tensions in the consortium. "Vivendi (then called CG Sat, part of the French water company

Company Générale des Eaux) were a partner, but they were vacillating. One day they were in, next day they were out. Finally we had to produce two complete sets of documentation each of which ran to many volumes – one with Vivendi in and one with Vivendi out." Eventually Vivendi did stay in, but with a smaller shareholding than MSI.

MSI had originally been down for a 5% stake of this but when one Egyptian shareholder defaulted, this rose to 7.5%. Suddenly the company had to raise $15 million and its senior investment partner Vodafone was expecting prompt payment. Kamiel Koot discovered that clause 6 in the articles of agreement allowed them to take 60 days to make their payment. Mo Ibrahim was a relative newcomer to the world of finance and for a small company without much of a track record it required strong nerves. He was terribly conscious that if he upset one of his most powerful business partners they might not do business again.

Despite these forebodings, the Egypt deal was closed on 1 June 1998 with $15 million from ING Bank against its shareholdings in India and Egypt and a $5 million equity subscription from the businessman Mo Ibrahim had worked with in India.

In 1999 Kamiel Koot approached Rick Beveridge to work on the process of putting the business plans together for investors. By the time he joined, the company had already acquired a number of licences. So Rick Beveridge's job was to write the business plans needed to attain new licences and the funding for them.

After the problems in Uganda, MSI had to explain why the majority of the revenues would be pre-paid. "The first month I was in the company I sat in a quiet corner and went through all the business models of all the licences we had. I then rewrote the business models and redid all the assumptions. We had to explain why 100 years of fixed line telecoms had not worked in Africa. The key changed assumption was that pre-pay mobile was the future of the business."

The team believed that it could also manage costs effectively and offset political risk: "We thought we could control costs and revenues. We will control political risk by getting a shareholder base (like the World Bank's IFC) which meant we didn't get beaten up by governments."

What the new business modelling told them was that if you grew slower then you might never make break even, whereas if you grew faster, the business would begin to require extremely large amounts of capital. The faster the growth, the more networks you had to roll out. In other words, the business needed to grow as fast as licences and money could be found, but there was a steady "upward suck" in terms of the scale of money required.

After Egypt and Uganda, the pace of growth became ferocious. As Kamiel Koot recalled: "The company was so cash hungry you just wouldn't believe it, we won licences left, right and centre. We had to roll out three networks: Zambia, Malawi and Congo B. This was followed by Gabon and DRC. There was even a point when the vendors were threatening to stop shipping equipment."

The next fundraising round produced $15.5 million in mid-1999 from a combination of the Worldtel Consortium (a fund set up under the ITU), $1 million from Mo Ibrahim and $1 million from Bessemer Investments, a US venture capital company.

The company may not have been making profits but the value of the shares rose at each fundraising round. The first investments were done on the basis of the shares being worth $2.06 and by the second round they were valued at $3.75 by ING – in the Worldtel round they were valued at $5.50.

However, it seemed like no sooner had this cash been raised than the company was out of money again by mid-1999. In February 2000 Citigroup came in to lead a round which raised $22.5 million.

But these cash needs were a sign that the company had found its market. As Terry Rhodes remembers it: "Africa was still not a crowded field in terms of the competition for mobile licences. Vodafone was going for big markets and were not interested in countries like Zambia. They preferred to expand in South Africa. So we started competing against the French and sometimes the South Africans."

Given the strong post-colonial ties of the francophone countries to France, it is still hard to believe that this was possible. But this is underestimating the scale of resentment felt in some of these countries for their former colonial masters.

As Rhodes tells it: "The Gabonese person (we dealt with) would say to us: 'The French ****** up our power, water and roads but they're not going to **** up our telecoms.' This was an opportunity for them (to do business) with a European player with good connections and an African focus. This made us the right international partner. We were European but we were not seen as the old colonial power and this was an attractive combination."

Whilst there were opportunities aplenty, finding money to fuel this growth was extremely difficult. The company talked to all of the donor-funded investment vehicles. The Dutch FMO was not interested at the time because it had had a previous bad investment experience in African telecoms. At this point, the World Bank's investment arm, the IFC, which had provided the money for the Uganda operation took the view that neither MSI nor Vodafone had acquitted themselves well. Luckily for MSI, CDC invested almost in spite of this early, rather faltering, track record.

More conventional bankers, however, avoided like the plague anyone who put Africa and telecoms in the same sentence. As a former banker himself, Lord Cairns (then with CDC) summed up the feelings of the banking community: "We were (seen as) slightly mad. Everyone knows that you'll never make money in Africa. They're all crooks and incompetents. There had also been a number of investments that didn't work out well. This was a feeling only reinforced by the fall-out from the telecoms and Internet boom.

"At CDC (later Actis), we took the view that it was worth backing because you had an entrepreneur who understood the technology requirements. It was a commendably honest operation and did not involve any underhand payments. And we had already made other telecoms investments in Africa and Bangladesh." Nevertheless CDC was taking a substantial risk. Its total African portfolio was between £300–400 million and Celtel was somewhere between 7–9% of the African portfolio – this was high in terms of a single investment.

So this interesting investment began to stretch even CDC's resources: "Every time we put money in we had to stretch our limit. We kept asking: can we put more money into it? Our whole African portfolio depended on it." Nevertheless the underlying pattern of performance gave cause for optimism: "It was a mixture of acquiring licences and rolling out network operation. It was growing faster than the cash flow coming in from different operations."

The impact of this almost daily struggle between growth and income made its mark on the company and cash flow over several years was extremely tight. Mo Ibrahim (and Kamiel Koot) sometimes had to cover the position: "Two or three times Kamiel phoned me to say that we weren't going to make the payroll. I covered this with a personal bank guarantee." As Kamiel Koot recalled: "Because the customer numbers were growing so fast, we always needed more money." Long nights were spent finishing the paperwork for the loans so that payments could be made on time.

During this time the company spread its financial wings and approached some of the major emerging market investors, including Mark Mobius of Templeton and Prince Al Waleed. Meeting the latter was a real experience as Kamiel Koot recalls: "Mo Ibrahim and I had to take a speed-boat to meet the prince on his yacht in Cannes. The waiting room was the size of a basketball pitch filled with the prince's trophies. After waiting a while we were presented to the prince in another palatial room. He wanted a substantial discount on the share price. Mo explained, in English and Arabic, that we had many institutional shareholders and could not offer such a deal no matter how cleverly it may have been structured. It was a pity because he liked the business. He gave us all his contact details for when he was in his plane, boat and tent but he did not invest."

The difficulty in persuading the financial community of the strong demand for mobiles hit the bottom line of the company hard. As Kamiel Koot recalled: "We thought the Congo-Brazzaville operation would cost $2 million. We looked at 5000 subscribers after year 1 and the banks said you're crazy to say that so we made it 3000 in year 1. There were 3000 subscribers in Brazzaville in week 1, so we had to find more money to invest."

During the period that followed, there was a repeating pattern: the company would win a licence and then struggle to make the payments. In Burkina Faso those responsible for the licensing process kept ringing up and saying: "Where's the payment? We said that we had SWIFT (bank) confirmation but we made a mistake on the SWIFT number to buy us time to get the $11 million. We kept saying to the Burkina Faso ministry the money is on the way. All the while (another competitor) was saying I may have put in a lower bid but I will match theirs."

In Chad it was the same problem. Even though the sum required was only $2 million, the company could not find it when it was due to be paid. Again Terry Rhodes remembers the same routine: "I started to get calls from Chad every few hours saying 'where's the money?' We had to stall for two weeks because we didn't have the money."

Worse still, the Celtel team had to present a large, ceremonial cardboard and foam cheque for the press cameras without it having been paid. There was a distinct feeling that the two people attending could easily have been detained for non-payment. But they actually left that night with the licence in their briefcases. However, the poor country manager for Chad was to find it difficult to get either money or attention immediately from a company that was simultaneously building another five networks.

Mo Ibrahim knew at quite an early stage that he would have to do an IPO soon. There were two pressures that moved him to this direction. The institutional shareholders wanted to "get liquidity" and the growing number of staff who had been given share options as a substitute for higher salaries wanted to realise the paper incentive they had been given. Therefore a proposal to float the company was put to its board as early as 2000 during the dot.com boom but it was decided it was too early to go to the market. And the discussion was kicked into touch when it acquired TTCL, as that company had no set of approved accounts.

Meanwhile their bidding rivals Orascom bought up another rival, Telecel, and launched their own IPO. But shortly after its IPO, the dot.com bubble burst and the share price fell. So instead of an IPO Celtel began to look at a trade sale and talks were held with Orange, Vodafone and Vivendi – the latter was judged the most appealing as it was offering $20 a share for 50% of the company.

By this time the company had sold both part of the shareholding in Egypt and the Indian shareholding for $12.5 million to British Telecom. Vivendi extended Celtel a $45 million loan, part of which was used to buy the Tanzanian incumbent telco TTCL. By 2001 Vivendi appeared to be moving seriously into Africa with its acquisition of Morocco's Maroc Telecom and the Kencell licence in Kenya. The drafting of documents for the sale to Vivendi took eight months and was due to close on 4 July 2001. Although Vivendi executives had already presented its credentials and plans to all the staff at Celtel's HQ, suddenly Vivendi stopped returning calls from Celtel and would not reply to its letters.

Vivendi itself was becoming a victim of the post dot.com boom. It said that it had found things of concern whilst carrying out due diligence on Celtel and did not want to proceed. The issues raised were whether Celtel had met its obligations under the licences it had acquired but these seemed like excuses for Vivendi, a company that was then in severe financial crisis.

In summer 2001, Celtel added to these woes by announcing it would claim damages from Vivendi and would therefore withhold its loan payments until the issue was settled in the courts. Under French case law if you drop out of exclusive negotiations, the other party can claim damages. So Celtel went to court and eventually won 500000 euros in damages for being the "jilted bride."

Aware of the worsening telecoms climate and the risk that the Vivendi deal might not be concluded, Celtel was simultaneously working on a plan B. It approached a consortium of institutions including Citigroup, IFC, CDC, DEG and Old Mutual for US$100 million in cash. But the terms of the deal got worse as discussions went on. IFC came up with idea of preference shares and a convertible loan with a guaranteed 20% rate of return. Under the scheme, it was possible to convert the loan into equity. The investors were clearly nervous and one by one each signed up to the proposal.

As one involved remembered it: "Thankfully, we closed the deal five days before 9/11. If we hadn't done that, we wouldn't be sitting here. We got **** from board members about how punitive the conditions were. But the alternative was bankruptcy."

This $100 million was used to pay back outstanding loans and to cover rising operational costs. After that, there was no money left to pay back Vivendi so Celtel took out another bank loan from Standard Bank and ING for $45 million. Fred Pichon recalled the rather bleak position after this loan was in place: "We went from cash hole to cash hole."

It might have sold its Egyptian stake but Celtel was having difficulty getting the money out of the country: "We had cash in Egyptian pounds that we couldn't transfer because of currency restrictions. There were all sorts of funny schemes to retrieve it and we took a huge haircut on that."

Up to this point Celtel had been, in the words of one person involved, a "projects organisation" and founder Mo Ibrahim was the glue that held the management team together. But he knew that it needed to be transformed from a start-up into a more conventional company with more formal processes. There were 60 people working at its headquarters with many more in the field.

In March 2001 board member Sir Alan Rudge was persuaded to become a part-time interim CEO (alongside his existing jobs in other companies which included a non-executive directorship of the ill-fated Marconi) and after a dose of the founder's legendary persuasive powers he reluctantly agreed. He was brought in to put in place a formal structure and processes and a human resources function. Sir Alan was a former deputy CEO at BT and was also brought in because it was thought his fixed line experience would be useful in making sense of the recent TTCL acquisition.

Mo Ibrahim knew what needed to be done: "When I started I needed to be both chair and CEO in the early stages to provide clear purpose and leadership and ensure discipline. I had to make sure there were no internal politics. There then comes a stage when the company becomes an institution. At that stage, you need to look at the structures again and lay down the basis to run the company. You need highly professional managers. Entrepreneurs are not good managers. I went through all this in my previous company and I understood that you must have good professionals."

The post dot.com boom period was a time when there was no money and little hope for the telecoms industry as it was widely seen as a victim of its hubris. Sir Alan had a very difficult balancing act to perform: keep the momentum but hold back from further acquisitions so the company could draw breath. He acquired a reputation as "the man who liked to say 'no,'" but in retrospect some of those involved understood why he took this stance and respected him for it. With the benefit of hindsight, one member of the team thought: "On this basis, he probably saved the company in 2001 because we didn't do any new expansions."

But the most difficult part of what he had to do was rein in the free-wheeling, informal atmosphere of the company that often valued speed of movement as much as the ability to take ground: the advance scouts were nearly always ahead of the main bulk of the army. But it was this "do or die," entrepreneurial spirit that motivated many of Mo Ibrahim's core team and they were suspicious of what they saw as the introduction of "bureaucracy." The fact that Sir Alan Rudge came from BT made it worse as in some minds he represented the old order they were trying to overturn.

One of Sir Alan's first moves was to introduce a new chief financial officer, David Wilson, who was also from BT. But as a sort of compromise, the financial responsibilities were split between David Wilson and Mo Ibrahim's unconventional banker Kamiel Koot, with the latter retaining responsibility for corporate finance. As Kamiel Koot rather dryly observed: "It was not a recipe for cohesion," and he was soon to move to take charge of its new payments company Celpay.

Since 2000 the company had had a central planning board that was there to approve all capital expenditure. This was originally set up to make sure that the company could get the benefit of centralised procurement, could share lessons in terms of good engineering practice and could make sure that all investments made sense from a financial point of view and met the internal hurdle rates.

In late 2001 this became more of a "capital rationing" board; Omari Issa, by then the COO, and his team performed a form of portfolio analysis on the operations and divided them into strategic assets that had to be kept alive at all costs – problem children that were costing the group money and had to be fixed, and a group of intermediate operations that were being well run but were under no significant threat and could be put "on hold" until times got better.

The result was that the companies that were "on hold" got hardly any investment and any cash that was generated by them was siphoned off and spent either on the problem children or in the strategic assets.

In order to maintain morale and momentum this was all done without explaining to the operations what was happening and all sorts of excuses were used to avoid investing. In Malawi, where the regulator had imposed a "Sender Keeps All" interconnect regime, Celtel publicly announced that it was stopping all investment.

With this approach the group became cash positive in May 2002 and those in the know were significantly more confident that the company would survive.

Although it was having a tough time raising funds, at the end of 2002 it was approached by MTN with a view to a merger. As one involved recalled: "The fit was perfect, like a hand in a glove." This was the third time the two companies had talked about this process but the first time that they had got to the due diligence stage. The attractions of the deal were enormous as both were largely present in different countries and together they would have made a formidable team.

There were many issues in play but what led to the negotiations breaking down from Celtel's point of view was the size of the percentages of ownership once the deal was complete. Tough talking led to one understanding (which actually got to signed heads of agreement) but MTN came back with changed percentages and by February 2003 neither Mo Ibrahim nor his board felt the deal was acceptable.

Sir Alan Rudge had finished his contract in December 2002 and in August 2003, the company appointed its first full-time CEO, Marten Pieters. Previously, when Mo Ibrahim had run the company, he had been both chairman and CEO and remained based in London, and Sir Alan was only involved for two days a week. But Marten was now full time and overseeing the operation from its headquarters building in Holland. He had come from the board of KPN (the largest Dutch telecoms operator) and was both used to the internal politics of large organisations and a very skilled people manager.

Marten Pieters agreed with the board that his tasks were three-fold:

1. To professionalise the company.
2. To grow the company even faster.
3. To give the owners an exit.

Celtel had already tried three times to sell the company. But at this point, the idea was to sell to a bigger operator. Mo Ibrahim had at the back of his mind that he would sell the company to someone like Vodafone but that did not happen. To all institutional investors, the company looked pretty tiny. It had revenues of $300 million but in the context of the global industry that made it a small company. But once you put it together with a bigger player it began to make sense.

So the company was to continue to grow but entered a period of calmer waters. The investment climate changed and after mid-2003 financing once again flowed back into the company. However, it became difficult to get the operations – used to having their investment plans rejected – actually to spend the finance that was now available. Within a year, it began to plan the IPO that was to lead to the sale of the company.

CHAPTER 13

Going to market – Celtel sells out

el point
celtel
G.T.L HARDWARE
DEALEARS OF
CEMENT, PLYWOOD, MABATI
AND E.T.C
Our wo
ONE SERVICES
GTL HARDWARE
NGUVU
80
KAH 703M

With its purchase of Kencell in Kenya, the company signalled its move up into the big league. Its transition to a more conventional management structure was giving it greater strength and credibility. Its new branding was carrying the message of a confident company capable of taking market share off its rivals in tight contests. The time had come to look again at how it raised its money to fuel its growth and it returned to the possibility of an IPO.

Celtel was no stranger to IPOs as it had previously prepared to float the company in 2000 just before the internet bubble burst. The impetus was two-fold: first, key employees had share allocations and were expecting at some point to be able to "cash-out" their paper rewards; and second, although the company was now much larger, the price tag of its acquisition targets was beginning to slip out of its reach. Even rival Vodacom's CEO Alan Knott-Craig was complaining that prices were going too high. Celtel needed access to the wider source of funds which the stock market provides.

But something always seemed to get in the way of reaching the finish line. An IPO is like lining up the cherries on a slot machine: all the different circumstances have to be right, and on previous occasions one of the cherries always failed to come into line. An acquisition would be too tempting to pass up and the plans would be put to one side. The market – as during the Internet and telecoms bust time – would turn against the idea of telecoms. But now the omens were auspicious and Celtel prepared itself for another attempt at an IPO.

In the autumn of 2004, it began a steady but insistent global press campaign announcing that it was shortly to float. The bare bones of the story were the same wherever it appeared. The company was "one of Africa's largest and fastest growing mobile phone companies" and it was preparing to float at a valuation of $2 billion on the London stock market: it would sell 25% of its shares.

The prospect of a simultaneous flotation on the Johannesburg Stock Exchange was thrown in for good measure because other African mobile companies were listed there. Celtel's chief strategy officer, Terry Rhodes, told the South African *Sunday Times* that the changes in South African law, which facilitated a secondary listing for African companies, "could have been written for us."

As Africa's third largest mobile phone company, it had revenues of $446 million in 2003 with operations in 13 African countries: it had just signed its five millionth customer in late 2004. This compared to market leader Vodacom, which had 13.5 million customers, but fewer than three million outside South Africa. It might not be the largest company but it was no longer a minnow. Potential share buyers were offered what in any other circumstances might sound like some highly unlikely scam: company growth over the last five years was some 1000% compared to a world average of 19% but average mobile penetration in the sub-Saharan African part of the continent was still only 3%. Unlikely as the figures sounded, city market investors now nodded wisely when the African mobile success story was mentioned rather than making their excuses and leaving.

Nevertheless the issue of political risk in Africa was still an underlying issue, as CEO Marten Pieters acknowledged: "The environments we're in are very tough. They aren't predictable. A coup in one nation may close down the network for two or three weeks and affect our quarterly revenues. Yet the governments in Celtel countries have never revoked a licence, so it would only be a short-term problem. Would the markets understand that? We can't set our targets too low, but if we're overoptimistic, we risk falling short. Will the markets understand Africa?" Certainly there were some concerns about whether a company such as Celtel, with its array of exotic African assets including the profitable but politically sensitive minority interest in Mobitel Sudan, was suited to the public markets.

Celtel press briefings pointed out that the company had raised $450 million in its six years of existence and the money raised in a listing would support future growth and also help repay the $250 million raised privately to acquire a majority stake in Kenyan operator Kencell. The line taken was that it had not yet talked to investors as it was finalising its results for the year to the end of December 2004 but the company was clearly semaphoring vigorously that it was an interesting and valuable purchase for prospective investors ahead of the more formal roadshow process.

Key elements of the IPO process began to fall into place. Shortly after the press campaign, Linklaters were recruited as legal advisers, Rothschilds as financial advisers and Citibank and Goldman Sachs as joint global coordinators and bookrunners. Lord Cairns joined the board as deputy chair at the beginning of 2005. He was appointed for a number of reasons but most importantly because he was a name known to the City and would be reassuring to them. However, as someone who had also been involved in one of the company's key shareholders, CDC Capital (later Actis), he was also well acquainted with the company's track record and with Africa.

As the company had anticipated, all this talk of an IPO had excited a number of potential trade buyers to step forward. As Lord Prior said: "We decided that the timing was about right (for the IPO). I was convinced that it would never come to an IPO. Once we announced the IPO, it would become clear we were in play and somebody would come forward." So before too long, the company had two parallel teams working on an IPO and a trade sale, both vying to be the one that came in with the best acquisition price. As Fred Pichon recalled: "We didn't want to end up in a situation like the one we had with Vivendi (where the buyer walked away)."

Celtel's IPO committee, along with the help of Linklaters and Rothschilds, structured the dual IPO and trade sale process to give the company the strongest negotiating hand. As Charlie Jacobs of Linklaters said: "by ensuring that the IPO was always deliverable we could keep the trade buyers 'honest' in the negotiations. If they sought too much in the way of warranties or indemnities, we could always respond by saying the company would rather do the IPO than a sale on those terms."

In order to make the trade sale easier, bidders were only invited in after full preparation of the IPO documentation; there was limited M&A documentation and the shortlist was kept down to three bidders: MTC Kuwait, MTN and one Asian telecoms company. Each was given a deadline of 21 March 2005, by which time each needed to submit a firm offer. There were two offers: one from MTN and the other from MTC Kuwait. The Asian company did not submit.

The board was quietly confident of the price but expectations rose as the day of reckoning drew closer. As Lord Cairns remembered: "At the time we thought the price would probably be between $1.8 and 2 billion. It went up as we went towards the IPO date. We thought the trade sale might be a bit more and the numbers went up as time went on. We all thought that something north of $2.7 billion would be acceptable."

Price was not the only relevant factor. With an IPO there would be new money for the company but only 25% of the shares would be sold. All the existing Celtel shareholders would then be "locked in" and unable to sell for a year. A trade sale would enable shareholders to cash in quicker. With MTN's offer on the table, it seemed that the long-time rival, which had already explored buying the company several times, would become the natural buyer. The advantages were enormous on both sides. There was almost no overlap in country operations so the resulting merged company would give a truly all-African operation. On this basis, cost and network savings would be tremendous. It also had the warm glow of leaving the whole operation in largely African hands, a considerable developmental plus for the continent. Indeed Lord Cairns had sold the idea to MTN's chair Cyril Ramaphosa on the basis that "you must do this for the good of Africa."

However, even the best dream ticket can be spoiled by other things. Each side was wary of the other because of the chunks taken out of each other in the course of Africa's mobile wars. You might protest publicly that it was all business and not personal but privately there were sore points. MTN might still have been sore that the Kencell acquisition had been snatched from under its nose. Celtel felt that it had been up to the altar with MTN before and that a deal was never done with them until it was legally documented and signed in all respects as the promised dowry ended up being smaller than expected.

By contrast, MTC's initial offer had lower numbers than MTN's $2.67 billion offer. Also it was a lot less exciting initially for the Celtel team because it was a pure trade investment. There was no great strategic fit between the two businesses. The Kuwaiti owners of MTC were looking for both long-term growth and returns.

So once again, Celtel's founder Mo Ibrahim and MTN's CEO Phuthuma Nhleko found themselves in a room negotiating the merger or takeover of the company, depending on which way you might spin the result. One of the key issues was that MTN wanted to close the office in Amsterdam but by Dutch law Celtel had first to consult the Celtel works council before agreeing to the sale.

Therefore the Celtel board could not approve the offer without finalising a number of points including: the division of management responsibilities; the package for employees who would be made redundant when the office closed; and whether the pending sale of Celtel's satellite subsidiary, Link Africa, should or should not be a condition of the deal. On the

Tuesday and Wednesday before Easter 2005, Mo Ibrahim was working with his negotiating counterpart Phuthuma Nhleko, negotiating these and other points.

In order to prepare for a speedy sale, and with the Easter weekend approaching, the company's shareholders were asked to sign irrevocable commitments to the MTN offer subject to finalising the negotiations. As Fred Pichon recalled: "We said to them can you please sign them and send them back. We'll keep them in escrow until the final points have been negotiated. On the Thursday night before Easter we had obtained 72% of acceptances." At this point MTN must have felt that the sale was "in the bag." Despite the Celtel team pressing for the outstanding points to be resolved and the documentation to be updated to reflect the last few days' negotiations, the MTN team decided to get the last plane back to Johannesburg on the Thursday evening for Easter and come back after the holiday to finalise the deal. On the Celtel side, there were still a number of concerns about whether this would be a "happy combination" and whether the outstanding points would once again be used for a late renegotiation on price.

On Easter Saturday events took a different turn. MTC clearly had second thoughts about its earlier offers and came back with a much higher number. It had noticed movement in MTN's share price and knew something must be moving. Despite the seemingly dismissive first offer, MTC did not want to miss the boat. Celtel's advisers said it was going to come back with an offer of $3.4 billion. The MTC team said they were coming to London with a knockout offer and would work all Easter weekend to get the deal done and legally documented before the stock markets opened on the following Tuesday.

But MTC were willing to negotiate to get the deal done. It made a commitment to keep open the Amsterdam office, keep the Celtel brand in use and leave the board in place. The Celtel works council was summoned and the new offer approved.

Board member Lord Prior felt it was an offer they couldn't refuse: "MTN had started to ask a lot more questions and the deal could not be consummated until they were answered. The Kuwaitis came along with a better offer. MTN's was $43 a share and MTC's was $56 per share. Someone said we thought $43 was a good deal but this is a 'no brainer.'"

By the evening of Easter Monday, the Celtel negotiating team had an agreed deal with MTC and a near agreed deal with MTN. The MTC offer was not only $700 million higher, but capable of signature that evening if the Celtel board approved it.

However, pleased though the board was, it did not immediately say yes but first consulted its London lawyers to see whether it could proceed with the MTC offer or whether it had made binding commitments to MTN. The answer from its lawyers was no. The irrevocable commitments had not become unconditional as the outstanding points had not been resolved, so there was no unconditional offer from MTN to accept. For their part, Celtel's only condition was that MTC's bankers would provide a guarantee that it had the money to make the purchase.

By Easter Monday MTC had come up with letters from its various banks guaranteeing the availability of funds at the agreed price and there was a contract for signature by the evening of the same day. The Celtel board approved the MTC offer on Easter Monday evening, the legal documentation was signed and the deal was then announced to the press on the Tuesday after the Easter Bank Holiday. The deal more than doubled MTC's subscriber base adding to its existing operations in Kuwait, Jordan, Lebanon, Iraq and Bahrain.

MTN was understandably incandescent and started legal action. Its first step was to start a disclosure process to get details of the irrevocables released and to consider injuncting the sale. It was clear that MTN felt that the irrevocables were the core of their case as it announced to the South African press that over 70% of the Celtel shareholders had accepted its offer.

But even before the disclosure process was complete, South African market analysts were beginning to ask why if the deal was completed on the Thursday before Easter had MTN not issued a cautionary notice to the Johannesburg Stock Exchange. As one broker put it: "If MTN did buy Celtel for more than $2.5 billion, surely that was a material transaction which should be announced to the shareholders and the stock market."

On 19 April 2005 the matter went to court in London and Judge Anthony Colman – who has seen the relevant confidential documents – refused permission for MTN to see the minutes of Celtel's board meetings, and said that MTN had "an apparently improbable basis" for claiming any breach of contract. The irrevocables clearly existed but they were given on conditions that were clearly stated in the e-mails disclosed to MTN and these conditions were not met. Put simply, the outstanding points on the deal were never resolved.

Shortly afterwards MTN said it had decided not to pursue the legal action it had instigated and would instead focus on other opportunities.

For founder Mo Ibrahim this was a sweet day and the second time he had sold a company. His 21% share of the purchase price was over $700 million. All the various staff members with shares and options also reaped their rewards, and in addition, the Board voted for an additional amount to be divided up between all the staff working in Africa who did not have any legal share ownership. This $18 million allocation averaged out at six months bonus per person across all the African operations. In all, the founder shareholders and staff netted $1.4 billion.

The other shareholders who netted $2 billion from the sale included: CDC (now Actis), Citigroup Inc.; EMP Global LLC, FMO, the Dutch government's private sector finance agency; Bessemer Venture Partners; the IFC, the commercial unit of the World Bank; Capital Group; and the Africa Infrastructure fund. In the case of the latter it realised $214 million from the sale. In the case of EMP Global it had more than quadrupled its initial investment in the company.

Tom Gibian, managing director of EMP and chief operating officer of the Africa Fund, expressed delight with the group's investment in Celtel, its first and largest investment in Africa.

But the failed deal with MTN remains one of the great "might-have-beens" for the African mobile industry and the question hovers as to why it did not work out. According to Celtel founder Mo Ibrahim: "It was always the most compelling story of synergy and compatibility and would have been a clear and obvious step. Mistrust had built up between the two companies over the years. We felt MTN were never willing to put a full price on the table and there was always a risk of late negotiating points. MTC showed that a deal is never done until it is signed and unconditional. This is business. Nevertheless you maintain your relationships at a personal level. There were no hard feelings from my side."

The money from MTC came in two tranches, the first one was 85% and the second two years later was the final 15%. On 14 May 2007 MTC paid the final payment of US$467 million to Celtel's shareholders. In just 10 years Africa had gone from being a place that would barely interest investors to one where a record deal of $3.4 billion had gone through almost uneventfully. The continent was really waking up from its long slumber and the energy and resources of the private sector were playing a key part in its renaissance.

CHAPTER 14

Meet the new owners

The new owners of Celtel were themselves regional players with an established base in the Middle East, but MTC Kuwait had global aspirations and hoped that its new purchase would go a long way towards making this a reality. Better still for Celtel, it admired the management and values of the company and worked hard to keep these in place during the transition period.

MTC had an unusual start in life for a company with aggressive global ambitions. It was started in 1983 as a joint venture between the private sector and the government of Kuwait. Its main purpose was to launch mobile telephony for Kuwaiti citizens and it was one of the earliest mobile companies in the world.

MTC had a monopoly on mobile telephony in Kuwait until 2000; when the government issued a second licence MTC experienced a sudden shock as it lost a large part of its market to the new entrant. The government of Kuwait decided to take action and it sold all but 24.5% of its share in MTC in 2002.

This was a fairly fundamental transformation for MTC. It went from being a government-owned investment vehicle to being 76% owned by 15000 private sector shareholders. The largest of these shareholders is a Kuwaiti conglomerate called Al Khorafi. It controls four out of the seven board seats whilst the government only retains nominating rights over one seat.

The current MTC managing director and deputy chairman, Dr Saad Al-Barrak, was recruited in 2002. He started his career in 1978 as an electrical design engineer for the Kuwait Pre-Fabricated Buildings Company. Then, after a time at the Kuwait Institute for Applied Technology as an instructor within the production department, he joined International Turnkey Systems (ITS) as a project engineer in 1983. He dealt with pre- and post-sales support for integrated information systems.

But it was with ITS that Dr Saad Al-Barrak's career really took off. In 1987, he became general manager with overall responsibility for general and business management. Under Dr Saad Al-Barrak's leadership, ITS increased its revenue base from $5 million in 1987 to $100 million in 2000, making ITS one of the leading integrated information systems companies in the Middle East.

He joined MTC in June 2002 and laid out the company's $3 \times 3 \times 3$ vision. In the first three years the company would acquire a regional presence in five countries. In the second three years it would aspire to be an international company with operations in Africa and Asia. It set itself the target of 10 companies with 10 million customers.

In the third three years it aspires to enter the top 10 of global phone companies alongside other large players like Vodafone, France Telecom and Telefonica. MTC's last three-year goal is that by the year 2011 it will have 110 million customers and have a $6 billion EBITDA.

Soon after Dr Saad Al-Barrak joined, MTC entered into a marketing partnership with Vodafone Group PLC, resulting in the launch of the MTC-Vodafone brand.

Dr Saad Al-Barrak

The end of 2002 saw MTC acquiring Fastlink of Jordan from Orascom, an acquisition that was up to that point one of the largest in the region. In April 2003, MTC-Vodafone was awarded the second GSM licence in Bahrain against stiff competition. In September 2003, Atheer Telecommunications – a consortium 50% owned by MTC and the remainder by private Iraqi investors – won the licence to start a GSM network in Southern Iraq.

MTC had known Celtel's founder Mo Ibrahim from the days when he had been at BT and also when he had been providing network consultancy through his first company. As Dr Al-Barrak put it: "I loved the Celtel story." He approached Mo Ibrahim and said why don't we partner? Mo Ibrahim responded that the company was preparing for an IPO so Dr Al-Barrak (along with a number of other companies) suggested a trade sale. As the previous chapter narrated, after a dramatic last minute offer, Celtel was bought by MTC.

At the time the press were sceptical that MTC had overpaid. But with hindsight Celtel has proved to be a bargain. MTC was extremely pleased with its new purchase for as Dr Al-Barrak said: "We knew it was very good. It had a large customer base. Its growth profile was high and it was making profits." Better still for MTC in terms of its stated $3 \times 3 \times 3$ vision, it catapulted the company into a different league immediately: it met its second three year targets and some.

But it had not just bought the existing portfolio of operations; it was in a position to take advantage of a couple of large growth opportunities that Celtel would have had difficulty in funding. MTC was easily able to finance two large-scale acquisitions – one in Sudan and the other in Nigeria – and a smaller one in Madagascar. These were purchases waiting to happen but among other things, the potential winner needed deep pockets. This was one attribute that MTC brought that made Celtel's presence more effective on the continent.

MTC left Celtel's management to continue in place for two years after the purchase because it believed that it was of an extremely high standard. When MTC reappointed the Celtel board after its acquisition, it kept key figures in place. Founder Mo Ibrahim remained the chair and Lord Cairns remained on the board together with eminent Tanzanian Salim Ahmed Salim, a former secretary-general of the Organisation of African Unity. Mo Ibrahim said at the time that the board was composed of "an excellent combination of international commercial expertise from Africa and the Middle East."

It is only from March 2007 that it has begun to integrate group functions across the company. So things like strategy, policy and marketing will no longer be present in both Celtel and MTC but the operations themselves will be left as they are. MTC now has 22 companies: 15 in Africa and seven in the Middle East. From the spread of operations in Africa and the Middle East, it was felt that it was well on the way to becoming the global company it aspired to be by 2011.

More radically than the merger of company functions, the newly emerging global company will be rebranded. MTC employed the same branding agency that had worked on the Celtel brand so successfully, FutureBrand. After an extensive global research exercise, it recommended Zain as the new brand name. In Arabic it means beautiful and good, and it has several positive meanings in other cultures.

The parent company was rebranded from MTC to Zain in September 2007 followed by the successful rebranding of four Middle East companies.

Tito Alai, who led the Celtel rebranding, has been retained in this capacity; thus closing the chapter on Celtel.

Zain remains interested in growing its operations both in Africa and elsewhere. Dr Saad Al-Barrak says the company will continue to "target the top 20 African economies. We will continue to be a pan-African company, closely focused on Africa."

Africa has a relevance to the Middle East that is not always easy to perceive if you're sitting in London or New York. As Dr Saad Al-Barrak tells it: "It's our backyard. In two years' time, it will have a population of one billion people. Our cultures are close and there are many economic and political ties. Integration is an important issue. Africa is often seen negatively as a place of corruption, diseases and political instability. It is these things that make it unattractive to the larger players but work to our advantage as we become a global company."

Part of that integration is about creating a communications infrastructure to link different markets. "We are not playing a game of numbers on the continent. We are looking at building an infrastructure to link up Africa to the rest of the world," he said. Zain's planned investment for 2007 was $500 million to leverage cross-border links with Celtel operations in Kenya, Uganda, Congo and Chad. It has already extended its East African "One Network" roaming scheme to 12 countries in Africa and into the Middle East. Even after a consortium led by Zain had bid an eye-watering $6.1 billion for the third mobile licence in Saudi Arabia in March 2007, Dr Saad Al-Barrak said: "If an opportunity arose tomorrow, we have $2 billion to back our interest."

He explained why the firm is concentrating on the African continent: "Africa offered higher growth prospects than Europe and was closer to the company's Kuwait base. We are able to compete much better here than in Europe, where the market is advanced and dominated by big players." He said players from emerging markets would find it difficult to penetrate the European market, even through equity partnerships.

Interestingly those Dr Al-Barrak sees Zain competing against in the struggle for global positioning are not companies like Vodafone and T-Mobile but companies that themselves have come out of emerging markets like MTN and Orascom. For Dr Al-Barrak wants Zain to become "the leading company in emerging markets worldwide." With some 50 million customers, soon to be united under the Zain brand, it is well on the way.

And where Dr Al-Barrak led, others have followed: for example, Etisalat of the United Arab Emirates made its own move in Africa continuing the flow of Gulf money into the region.

Although Zain comes from a very different culture – both in terms of language and organisationally – there is an overlap in terms of how it sees its values. Indeed Dr Al-Barrak stressed that one of the attractions of buying Celtel was this closeness in terms of values.

Dr Al-Barrak added: "The essential cornerstones of our strategy revolve around excellence, diversity and benevolence. These values permeate everything we do and they are led by our commitment to our human resources that are our real treasure. We are fully committed to the development of our people and we are committed to ensuring that their incentives are aligned with shareholder objectives. We are copying the Celtel model by introducing a form of employee stock option plan. Our products, services and technology are tailored to meet our customers' needs and demands. Ensuring that we address all stakeholders is complemented by our commitment to our corporate social responsibility through the multiple activities sponsored by Zain and our adherence to best practices in all our activities and commitments."

Indeed Zain has seen its role in a very similar way to Celtel's brand slogan "Making Life Better." In 2006 it released an independently commissioned report called "Mobility for One Language, Diverse Cultures – The socio-economic impact of mobile phones in the Arab world" at the 3GSM World Congress in Barcelona. It highlighted both economic and social effects of mobile communications usage on the MENA region, developmental effects that were remarkably similar in some ways to those found in Africa.

At the report's launch Dr Al-Barrak said: "The report shows the potential of mobile communications to enhance not only the economic position of countries and people but also to change the social fabric of our communities. At Zain, we have always believed that mobile communications are part of the framework of societies and the daily lives of people and this report attests to that fact."

With this common understanding of how mobile phones can make life better for people across both Africa and the Middle East, Zain feels that it is well on the way to reaching its aspiration of becoming a global player that understands emerging markets. Whereas once multinational companies were based in Europe or America and sold developed world products to developing countries, the next generation of truly global companies seems to be coming out of an understanding of how to sell products and services to customers in emerging markets. The world looks like being well and truly turned upside down.

CHAPTER 15

Buying Nigeria's V-Mobile – the billion dollar deal

Celtel lost out in the bidding battle to get into Nigeria in 2001 but it returned to the fray in 2005 after it had been acquired by MTC/Zain. It knew that Nigeria was going to be the continent's largest market and that it was essential that it found a way in. The continuing legal and shareholder battles of one of the operators gave it the opportunity it was looking for.

The operator in question was Econet Wireless Nigeria which later became V-Mobile. Because of disputes between its international sponsor Strive Masiyiwa and its local Nigerian shareholders over cash and investment shortages, the latter were casting around for a new owner and investor to replace him. However, not only had there been disputes between the investors and Masiyiwa but also between the investors themselves.

The impact of these disputes had a clear impact on its competitiveness and its ability to meet the rocketing growth in the market. Its switches were full and it had little or no money to expand its network. A company source told a local reporter: "Econet needs cash to grow its network and meet the demands of the Nigerian market."

In August 2002, local paper *This Day* reported that Celtel was discussing acquiring the company and putting in a package of equity and loans, totalling hundreds of millions of dollars. In fact, this was the second time there had been talks as they had taken place a year or so earlier. But Celtel's bid for Tanzania had gone ahead and the company was struggling financially and did not immediately have the money to continue the discussions. In the event, this second round of discussions never really took off.

Meanwhile Celtel was also talking to the state-owned fixed line company Nitel about the possibility of getting involved in investing in its mobile network after Nitel stopped a similar discussion with Vodacom. David Easum, Celtel's business development director, told the *This Day* reporter that "they believed that a deal might come up in due course." However, these discussions were designed as a smokescreen for the true target which was to remain Econet Wireless Nigeria even though the discussions had gone cold.

Shortly thereafter Vodacom came back into the picture and also started pursuing Econet Wireless Nigeria as the more promising target. It first intended to buy a majority stake in Econet Nigeria in a deal worth some $150 million in equity and another $100 million in operational loans.

But Strive Masiyiwa's Econet Wireless International contested the deal in the Nigerian High Court. It sued Vodacom for what it described as "inducement to breach of contract," based on the fact that it had first option to buy if the local shareholders wanted to sell their stakes. Whilst both Vodacom and the local shareholders sought a legal strategy to overcome this obstacle, the Nigerian shareholders took the management contract for operating the company away from Econet Wireless International and gave it to Vodacom. The company name was changed to Vee Networks, everything was rebranded Vodacom Nigeria and Vodacom appointed its own CEO to run it.

Suddenly, two months later, at its results announcement in South Africa, Vodacom dropped the bombshell that it was pulling out of the deal. It had lasted eight weeks, and was described as the shortest corporate pact in Nigerian business history.

According to the *Financial Times* (London), Vodacom's surprise decision to pull out of Nigeria was due to a "breach of trust" between Africa's largest mobile phone operator and its Nigerian partner. Vodacom's chief executive Alan Knott-Craig said: "the decision to withdraw from Nigeria is a blow to our expansion plans, but there was a breach of trust and I decided we could not expose the company any further." He also added "those Nigerians are too smart for us."

The "breach of trust" was alleged to centre on the company's decision in March to approve payments totalling just under $3 million to three brokerage companies. Vee Nigeria directors held directorships in two of the companies. Vodacom had conducted extensive due diligence and found that earlier payments, which were negotiated in 2002, to brokers that had helped the company raise capital from Nigerian state governments were not illegal.

However, Knott-Craig said he specifically requested that no more money be paid out. On 22 March, just over a week before the management agreement with Vodacom started, Oba Otudeko, Vee's chairman, approved the later payments, according to a letter seen by the *Financial Times*.

"We had a clear understanding, but Vee chose to pay without telling us," Knott-Craig said. "When we discovered they had signed the management deal after making those payments, it was the trigger for us to leave Nigeria. We shall not be going back there for the foreseeable future."

Alan Knott-Craig said earlier this year that two of its shareholders – Vodafone, which owns a 35% stake, and SBC Communications, which had an indirect stake in Vodacom through its shareholding in Telkom South Africa, which in turn owned 50% of Vodacom – were concerned about compliance with the US Foreign Corrupt Practices Act. The law, which threatens fines and jail terms, is designed to stop US multinationals and corporates and those companies with business in the USA from engaging in corrupt practices. "If it were not for that, we would be operating in Nigeria now," Knott-Craig told Moneyweb.

Vodacom's deputy CEO Andrew Mthembu, who headed its international operations, had his contract terminated and strategy director Robert Pasley resigned.

The *Economist* magazine captured the mood with a story on 3 June 2004 entitled "Vodacom, Vodago." "Maybe it was an April fool's joke. On April 1st Vodacom, a large South African telecoms firm, proudly announced a five year deal to manage Econet Wireless Nigeria … But this week Vodacom beat a hasty and not very dignified retreat."

Vodacom Nigeria was given three months to change its name, and so, in what was to the consumer the second name change in four months, the brand became V-Mobile.

Willem Swart, put in as CEO, also resigned from Vodacom, but remained with Vee Networks as its CEO. Vee Networks' public relations manager, Emeka Opara, said at the time that "Vodacom had last week issued a statement in South Africa to the effect that it did not find any evidence of corruption in Vee Networks after a thorough due diligence."

It was also agreed that Vodacom, in all its correspondences and statements about the pullout from Nigeria, would desist from alleging that it was because of corruption as Vee Networks insisted the brokerage payments were not bribes.

So Vee Networks, as the company was now known, was now stranded without a convincing investor. Worse still from a potential investor's point of view, the local shareholders were still in long-standing litigation with Econet Wireless International.

The local investors started once again in their efforts to find a buyer and to resolve how they might undo the legal knot at the centre of any potential transaction.

A year later the race was rerun with a slightly more extended field. One of Vodacom's key shareholders, Telkom South Africa, had its US shareholder SBC sell out, enabling Vodacom to return to the battle with a slightly freer hand, though it still had Vodafone looking over its shoulder. Richard Branson's Virgin began to express interest in buying the company. Its strategy was to enter major markets and leverage the profile of its brand. Virgin Atlantic was flying into Lagos and it had proved very successful so the company was investigating whether it might build on this initial success. In February 2005, Virgin was reported to be looking very closely at the company and a decision was said to be imminent. By spring of that year, both Vodacom and Virgin had made separate offers but eventually decided that the two of them would bid together. However, for reasons which are still not clear, something happened and Vodacom and Virgin broke off negotiations.

Meanwhile Celtel were still hoping to be in the race, but from October 2004 the shutters had come down on all such acquisitions as part of the strict legal process running up to a London Stock Exchange listing.

After the MTC/Zain acquisition of Celtel that constraint was removed, but the situation was still complex in Nigeria. Celtel's Dave Hagedorn had the unenviable task of running another feint offer for state-owned fixed line operator Nitel: "The backstop position was that we might have to try and end up with their mobile operation. We signed up for the full expression of interest (to buy Nitel) and had access to the data room. It was a disaster. The last audited accounts were from 2003.

"We went to a management presentation and I have never seen anything funnier. Nitel gave completely different numbers in the morning to those given by Mtel (its mobile subsidiary) in the afternoon. Apparently one set of numbers had been put together by the privatisation agency, BPE. There were two occasions during the question and answer session when the audience just laughed at the Nitel management. It came to nothing but it was a useful way of keeping the pressure on the V-Mobile shareholders."

A new round of discussions opened up and Celtel rejoined the dance. Both Vodacom and Celtel made fresh offers and the local Nigerian shareholders agreed to pursue the Vodacom offer. However, by January 2006, Celtel was back in serious discussions with the company's Nigerian shareholders and the following month the local press reported that Vodacom had once again withdrawn from negotiations.

So Celtel had a clear run at the company but it faced formidable obstacles. It knew that Econet's founder Strive Masiyiwa would be watching like a hawk to see how they dealt with the shareholders' agreement and would not hesitate to take legal action. The sale was to be handled by V-Mobile's Investment Committee that represented the interests of all the local shareholders.

The strategy devised to undo the pre-emption knot was simplicity itself but required deeper pockets than Celtel now possessed since its purchase by MTC/Zain. It would make an offer to buy the shares of all shareholders and by doing so it would trigger the pre-emption rights Econet said it was entitled to. Once this had happened, Econet would be given the opportunity to buy all of the company if it could match the Celtel deal – it was in effect a case of "may the best man win." But it also had the virtue of limiting the period during which the shares might be acquired to 30 days. So not only did it require a lot of money to be successful but it had to be done quickly. For other litigants, escrow arrangements would be created to ensure that if their claims were successful and court decisions were decided in their favour they would not be prejudiced.

The local shareholders of V-Mobile wanted the certainty of completing the deal because almost since its inception the company had always been in a dispute with either Econet or other local shareholders. They wanted to put an end to the litigation to take some of the profit and start focusing on the business.

So after years of successive legal actions, Econet now had the opportunity to buy the company as it always said it wished to do. On 15 April 2006 Celtel made an offer to buy the shares of the local shareholders of V-Mobile for $755 million together with shareholders' put options worth about $460 million, a total amounting to more than $1.2 billion. In addition it said it would add to V-Mobile's share capital by subscribing for 32.89 million shares for a total consideration of $250 million.

Econet Wireless International, in accordance with the Shareholders' Agreement, were to be given the opportunity to match that offer and, by an offer letter dated 15 April 2006, it was informed that the Nigerian shareholders were minded to accept Celtel's offer but it was being given the right of first refusal in respect of their shares. It then had 15 days in which to accept this offer; if it had accepted another 30 days before it would have had to credit the account of First Bank of Nigeria PLC, who were acting as escrow agents for the transaction, with the sum of $1215823497.60, failing which the offer "would be deemed void and of no effect."

Push had at last come to shove. Econet Wireless International failed to meet the timetable and when it asked for an extension, V-Mobile refused it. Econet then headed to the courts in London to seek an injunction restraining the transaction with Celtel. Initially the injunction was granted without V-Mobile or Celtel being present, but then on its review Econet's luck began to run out.

Justice Morrison hearing the review said that the governing law for the Shareholders' Agreement "was expressly stated to be Nigerian Federal Law," nevertheless the injunction was sought by Econet in the English courts. He then went on to deliver a damning verdict.

He said that: "On 2 May 2006 Econet purported to accept, uncondi-tionally, the offer of pre-emption." In other words if Econet could match or exceed Celtel's offer, the company would be in its hands. On 9 May Econet's Nigerian lawyers received the transaction documents with confir-mation that the deadline for payment was 18 May 2006.

Econet wanted an extension of time because it said that the transac-tion documents were not acceptable to Econet and its financiers. However, in an application for an injunction to the English courts on 13 May 2006, its lawyer (basing his evidence on his own knowledge or "supplied to me by the source stated") said: "The Applicant (Econet) has access to the funds to enable it to make the payment of $1215823497.60 subject to obtaining the Transaction Documents mutatis mutandis in executed form and subject to the appointment of an agreed escrow agent can then make payment."

But as Justice Morrison stated in his judgement: "But even taking his figures at best, and ignoring qualifications and pre-conditions, on the figures, the total amount was $90 million short. But realistically the equity funding (supposedly for $400 million) was not committed but rather was offered subject to conditions; and of the debt funding ($975 millions) there was a contractual commitment as to part and no contractual commitment as to $475 million."

All this led Justice Morrison to the unequivocal statement that: "Accordingly, the bald statement that as at 15 May Econet 'has access to the funds to enable it to make the payments' subject to the completion of the transaction documents and agreement as to the identity of an escrow agent was not true." He also emphasised that: "It was also a statement that was not made on the basis of the deponent's (the lawyer's) own knowledge." The statement was made on the basis of evidence supplied by Econet but it was not stated to be the source and "a disclosure application had to be made to Coleman J. to obtain it." Put simply, when Econet was asked to find the purchase price offered, it was unable to do so.

The way was now open for Celtel to complete the transaction. The acquisition team tumbled out of the court as quickly as possible and headed in a taxi for Heathrow. They almost missed their Virgin Atlantic flight to Lagos but were to arrive in time the next day to close the deal with the company.

When the acquisition was announced to the Nigerian press, CEO Marten Pieters stressed that it was going to rebrand the company as Celtel but because of the scale of the task, there would be a delay before this happened. It would use $250 million of the acquisition money to expand the network and thereafter spend $300 million a year on growing it even further.

Even before this announcement, Econet had announced that it was going back to court to challenge the acquisition. But as Pieters noted: "When we negotiated the acquisition we gave Econet the full pre-emptive rights, although the shareholders in Nigeria deny it is a shareholder.

"We didn't want to run the risk that after two years in the courts Econet would win the right to those shares. So we offered them the rights and if they had come up with the money it would have been game over for us. Econet wanted to buy the stake, but failed to raise the cash."

Econet sought an urgent interdict in Lagos to force V-Mobile's shareholders to release the documents covering the pre-emption rights. Strive Masiyiwa said he was confident that the court would find in Econet's favour, and reverse any attempt to sell the shares to Celtel. "If they buy disputed shares, they must know that they are holding them in trust, as those are our shares."

But Celtel had the upper hand and had acquired a controlling stake in the company. As communications director Martin De Koning told *Business Day* on 2 June 2006: "We are an operator that operates; we don't operate in the courts."

Business Day's article on 2 June 2006 identified the state of the loser: "Much as Masiyiwa will object to this, there are times when he seems to spend more on lawyers than he does on technicians. Just ask JSE-listed Altech, which thought it was a smart idea to form a joint venture with Econet to become a cellular operator. A year down the line the deal collapsed, with the companies practically talking only through their lawyers after failing to reach an amicable annulment.

"Econet has also instigated legal action against Vodacom, when that operator made a play for V-Mobile. In the meantime, V-Mobile has suffered because nobody would invest while its share structure was under dispute … Econet has deservedly gained a reputation as a litigious entity, which is a dangerous thing to acquire and a difficult one to shake. At some stage – and perhaps it is crossing that threshold now – investors, potential partners or potential acquisition targets that see a good business case in Econet will stay away because the baggage that comes with it is not worth taking on."

On 10 June 2006 the Lagos High Court dismissed an application for an injunction to prevent the country's regulator NCC from giving approval for the sale. However, various other actions continue in the Nigerian courts and are likely to rumble on for some years. But by September 2006 the company had changed all of its branding to Celtel and was beginning to gain market share.

As Celtel board member Lord Cairns observed: "The Nigerian deal was an interesting one. From an income perspective, it was a high priority acquisition. It was going to be a lot of money and that had to come via MTC/Zain. With them now taking the risk, there was a reasonably high probability of it working out well."

To put it in context when Celtel was bidding for the original licences in Nigeria in January 2001 it had 170000 customers in nine countries. Five years later, the acquisition of V-Mobile took Celtel's total customer base to 17 million across 15 African countries. In late 2007 Nigeria overtook South Africa as the largest market on the continent and Celtel celebrated its 10 millionth customer in Nigeria alone.

CHAPTER 16

Luck or good judgement?

Some of Celtel's successes were the outcome of careful analysis and learning the lessons of failure. Others were the result of good luck: either being in the right place at the right time or not being in the wrong place at the wrong time.

In particular Celtel was lucky with the order in which it met its major competitors – it came up against MTN early, and learnt the lessons of pre-pay and only came up against Vodacom at a time when it was ready to learn the lessons of having a strong brand. Otherwise the timing of external events was, in some ways, fortuitous – the end of the dot.com boom only came at a time when the company was able to get itself cash positive and wasn't so financially overstretched that, like Vivendi, it had to sell the family silver. If it had come earlier it would have been more problematic; if later its competitors may not have been so hard hit.

Celtel, like Napoleon's best general, was lucky. But it would be wrong to imply that luck alone built the company, for it triumphed over considerable odds and did so because it was prepared to do things differently. It was always aware of the importance of reputation. Warren Buffet said: "It takes 20 years to build a reputation and five minutes to lose it. If you think about that, you will do things differently."

Was Celtel a Black Swan? In Nassim Nicholas Taleb's definition "a rarity, with extreme impact and retrospective (though not prospective) predictability." Western perception was that all swans were white, until the discovery in the seventeenth century of black swans in Australia. Was there a similar Western perception about African business success before Celtel?

With its origins in MSI, it was both well connected to institutional investors and to some of the key "movers and shakers" in the telecoms industry at a global level. Companies like General Atlantic Partners added credibility to what MSI Cellular was doing even if at this stage it was not the activity for which they invested in the company.

Through selling consultancy and software to the first and second generation of new mobile companies, Mo Ibrahim sat at the same tables as the senior managers of these companies. This enabled him to build a high profile board that included Sir Gerry Whent, who was the first CEO at Vodafone, and Jay Metcalfe, who had been the CEO at Millicom. In addition Mo Ibrahim had recruited a senior US venture capitalist, Felda Hardymon of Bessemer Ventures and Harvard Business School.

Through these connections, the company had access to the "brightest and the best" in the industry. But by themselves, all these connections would have counted for nothing without founder Mo Ibrahim's legendary powers of persuasion. Early in the life of the company, Mo Ibrahim was able to persuade companies like Vodafone that it had something to contribute in Uganda and Egypt.

With MSI and later with Celtel, the focus of interest was initially at a global level. There were deals to be found everywhere and indeed two of the first three of Celtel's shareholdings were in Asia not Africa. But through a mix of factors, Africa emerged as the chosen opportunity. Ibrahim enjoyed working in developing countries and as an African it is hardly surprising that he thought hard about his home continent.

But with hindsight, Africa also came into view for another important but less well acknowledged reason. It was one of the few places left on the planet where licences were still obtainable at a price that could be afforded by a small company. If you were prepared to operate in places that were seen as risky, like countries emerging from civil wars, you could scoop the prizes for your daring. But the strategy of collecting what might be described as the "low cards in the pack" would not by itself have allowed Celtel to emerge as one of the largest regional players. Later attempted (Nigeria) and successful (Kenya) acquisitions were of such a scale financially that without its determination and support from its investors, Celtel could easily have languished in the second division.

It is hard to remember back to the days when mobile licences in Africa were not really anything to write home about. A few thousand subscribers paying very heavily for the privilege of their new phones produced a nice little business but were not of sufficient scale to attract anyone's sustained attention. MTN had not entered the Ugandan market with such striking results. Orascom had bought Telecel but failed to integrate its management and seemed almost ambivalent about committing itself to these new markets in sub-Saharan Africa.

It is also worth remembering that Celtel was not the "first mover." The most significant "first mover" was Telecel and it could so easily have taken up the position Celtel was to fill if it had played its cards differently. But from the beginning Telecel was a very different kind of operation and a comparison gives some idea why it didn't succeed.

Rwandan-born engineer Miko Rwayitare teamed up with an American pilot called Joe Gatt to launch Africa's first mobile operation in DRC in 1987. It had a tax avoidance structure involving both owners having their own offshore companies and an operating office in Reston, Virginia. Although the analogue system was expensive to run and all subscriptions are post-paid, both the phones and calls were very expensive so the company was making a lot of money.

But it failed to innovate technically and neither of the owners seemed very interested in raising financing through the usual routes or in building anything resembling a conventional management structure. It relied heavily on the dual personalities of its owners and once they fell out, the company was all but finished. Neither had much sense of a bigger vision and seemed primarily interested in using their "little oil wells" for their own purposes.

Ironically a number of key Telecel personnel joined Celtel when the company collapsed, bringing with them much needed experience of working in Africa in the mobile field. As one of them, Fred Pichon, observed: "It failed to consolidate itself. The main aim of the two owners was to take money (for themselves) out of the business. It had only really one good business and that was in DRC. Mo had a much better vision."

In many ways they represented the old Africa where companies took lucrative franchises in business, without there being any impetus to change business or technology practices. As one of the people who watched the operation from within Celtel observed: "When equipping their network, the question was always which supplier will offer the best financing deal today? As a result it had a mish-mash of equipment."

When a clash of personalities between the two owners led to the collapse of the company, Celtel bid to acquire it but the sale dragged on as neither of its two owners could agree between themselves. Eventually the assets were separated and in March 2001 Naguib Sawiris' Orascom Telecom bought most of the operations for $413 million after he had failed to persuade Mo Ibrahim to fold Celtel into Orascom ahead of Orascom's IPO.

Ironically, given their different philosophies, Celtel and Telecel had tried to team up in June 1999 to bid for the third licence in South Africa. The black empowerment element of the bidding consortia was a key factor, and together Mo Ibrahim and Miko Rwayitare were the predominant players in sub-Saharan Africa. They appeared together at the public hearings in October 1999. The winner was supposed to be announced by December 1999, but it was not until November 2001, after legal and administrative wrangles, that the licence was eventually awarded to Cell C, led by Saudi Oger, a company with a minimal track record of cellular operations.

Mo Ibrahim and Miko Rwayitare was a marriage of convenience, and in hindsight it was probably a lucky escape. It allowed Mo Ibrahim to focus on his own company and to inspire his team through a combination of vision and passion and of sharing the wealth the company created with them through share options. However hard or difficult things became, most of the key people could see that the share value of the company was rising and one day they themselves would gain part of the value they had created.

Companies, especially multinational companies, talk about having vision but more often than not the "vision thing" they are talking about is a formulation of banal words that do not necessarily command respect or attention within the company.

But Celtel's vision combined an "inclusive capitalism" with a strong sense of idealism. Africa might be one of the hardest places on the planet to get things done but if you actually succeeded in doing so against the odds, the sense of achievement was overwhelming. Better still, the customers were positively gagging to get their hands on your product to the point where hundreds of them would hammer down your doors in the attempt. This is not a feeling that the person selling widgets on a wet Wednesday must often experience.

As a result, there was a strong sense of ownership that survived the occasional harsh argument and tactical disagreement. All staff put in extraordinary hours to get things done, particularly during the difficult years when money was short. Working overnight on loan documents would mean that others would get paid on time. As Fred Pichon remembered: "It was a fantastic experience."

But it was neither financial reward nor a strong sense of ownership that glued disparate staff members together. It was a sense of idealism that they were creating something that would not only make money but change African lives forever. At an individual level, Senegalese Mamadou Kolade did what he did because he wanted to show that Africans could actually do business properly. For Cape Verdean Emily Macauley, every working day would bring sights and stories that would show her for the first time in decades that Africa was on the move and lives for some its poorest were changing before her very eyes.

"Making Life Better" might be an advertising slogan but it encapsulated so much of what the company was achieving. Without any cynicism, the company was able to take the narrative story of improving lives and use it for PR purposes. Almost for the first time, the private sector was not the "robber baron" figure pillaging the country's natural resources but a force of nature that was delivering things that African politicians could only make speeches about.

Liberalisation of African economies had until the arrival of the mobile companies been a still-born affair, heavily contested by some Africans and many campaigning NGOs. With the arrival of the mobile operators, the sheer scale of investment produced almost immediate results at every level. Africans got good jobs, were well paid and had their aspirations raised.

Advertising "Making Life Better" with aspirational black images signalled a different road from the slow progress made when dependent on donors. To be fair, some of the donors learnt this lesson and backed Celtel. Not many managers elsewhere could go home at the end of the day saying that they had arranged payments for soldiers going home at the end of the civil war in DRC.

At the heart of this idealism were a set of ethical standards that Mo Ibrahim had learnt from his life in the UK and he was determined that the company would stick to in Africa. But his attitude was both highly moralistic and pragmatic. He often said corruption was like adultery – it took two parties: the bribe giver and the bribe taker. He believed that if Africa could do business in a different way it would attract new international investment from people that did not want to take part in the old ways of Africa.

The Celtel story is not one of seamless progress towards its current shape and size. The speed of growth meant that often whilst things got done, they were not always done well. As one of the team in the early days remembered it: "We weren't very good at gauging the level of demand. It generally took us by surprise. It took everyone by surprise. We thought we'd be dreaming if we achieved 2.5% penetration and certainly the banks would never believe such figures were achievable. In the end, the figures were much higher. Initially, we didn't have the people, neither at an international nor at a local level. It took us a while to gear up at both these levels."

As it scrabbled to keep up with the rocket-like ascent of the organisation, it did not always find the right people. To be fair, it was looking for large numbers of people prepared to work in Africa, preferably with some telecoms experience, and this was not always a large pool of people. Particularly in the early days it had to attract the kind of person who metaphorically was prepared to wrestle snakes and in reality might stay for several months in a hotel without running water or windows.

Nevertheless as it grew to be a substantial company, it empowered Africans in a way few things before it had: over 98% of its staff were Africans. It took Africans from one country and put them in charge of things in another country, consciously wanting to break down the old bonds of obligation and favouritism that plague African productivity. Mo Ibrahim believed that diversity was essential to avoid a "me-too philosophy." And that you did this by widening the "gene pool" with an extremely mixed group of people who would challenge received wisdom.

Staff had to apply for their jobs irrespective of connections and were appointed for their professional skills rather than their family connections. It created 6000 high-quality well-paid jobs and there were some 40000 people dependent on it for indirect employment. It trained its top management teams at London Business School. This was capacity building on a scale and of a kind that donors could only dream about. In June 2005 when the company was sold every employee got a sale bonus that was an average of six months' pay.

But despite all these good things, the company scraped through in the early years and could easily have hit the wall. Nevertheless despite the considerable level of pressing financial demands, Mo Ibrahim knew that he had to take the company from a highly flexible "guerrilla army" start-up to a "standing army" that would take and hold ground as Africa became increasingly competitive. Again it is a tribute to Ibrahim's abilities at handling people that this transition was made with very little fall-out.

Celtel grew out of very particular circumstances but these may return as Africa liberalises more sectors of its economy. As a company that combined being absolutely African with being headquartered in Holland, its success is an example of bringing together cultures, technology and finance to drive development in Africa. If only there were many more such success stories...

ANNEX 1
Chronology of events

1989	Mohamed Ibrahim founds his own consulting firm, Mobile Systems International (MSI)
DECEMBER 1995	Celtel launches its mobile network in Uganda
MARCH 1998	Celtel is spun out of MSI and founded in Amsterdam under the name MSI Cellular Investments
DECEMBER 1998	MSI launches its mobile network in Zambia
OCTOBER 1999	MSI launches its mobile network in Malawi
DECEMBER 1999	MSI launches its mobile network in Congo (Brazzaville)
JULY 2000	MSI launches its mobile network in Gabon
SEPTEMBER 2000	MSI launches its mobile network in Sierra Leone
OCTOBER 2000	MSl launches its mobile network in Chad
DECEMBER 2000	MSI launches its mobile network in DRC
JANUARY 2001	MSI launches its mobile network in Burkina Faso
FEBRUARY 2001	MSI acquires 35% of TTCL in Tanzania
MARCH 2001	MSI acquires 39% of Mobitel in Sudan
JUNE 2001	Sir Alan Rudge appointed as interim CEO
NOVEMBER 2001	MSI launches its mobile networks in Niger and Tanzania
FEBRUARY 2002	MSI acquires LinkAfrica, an international satellite link supplier
DECEMBER 2002	Celtel launches Celpay in Zambia, a pilot system for m-commerce
DECEMBER 2002	Sir Alan Rudge resigns as interim CEO but remains a non-executive director
JULY 2003	Marten Pieters appointed CEO of Celtel
NOVEMBER 2003	Celtel is awarded the SMO prize by the Foundation for Business and Society for its Dutch investments abroad
JANUARY 2004	Celtel brand is relaunched and the holding company is renamed Celtel International

JANUARY 2004	Celtel secures a US$62 million funding via Capital International, a subsidiary of Capital Group
APRIL 2004	Celtel acquires 60% of Kencell in Kenya for $250 million
OCTOBER 2004	Celtel wins the first IFC Client Leadership Award for the contribution made to sustainable development
DECEMBER 2004	Celtel signs up its 5 millionth customer
JANUARY 2005	Celtel announces its intention to float on the London Stock Exchange, advised by Rothschilds, Citicorp, Goldman Sachs and Linklaters
MAY 2005	MTC Kuwait acquires Celtel for $3.4 billion; 85% now and 15% in two years' time
FEBRUARY 2006	MTC announces that it has successfully acquired 61% of Mobitel from Sudatel in a deal valued at US$1.332 billion to add to the 39% already owned by Celtel
MAY 2006	Celtel announces that it had reached an agreement to acquire a controlling stake of 65% in V-Mobile, one of Nigeria's mobile telecom operators, for US$1.005 billion
SEPTEMBER 2006	Celtel launches One Network, the first ever borderless mobile network in the world. This allows East African customers to move freely across geographic borders without roaming call surcharges and without having to pay to receive incoming calls
SEPTEMBER 2006	Celtel operates in 14 countries serving more than 15 million customers
DECEMBER 2006	Celtel CEO Marten Pieters steps down; Moez Daya takes on role as MTC Group's interim CEO for Africa
MAY 2007	MTC acquires the remaining 15% shares in Celtel for $467 million
JUNE 2007	Celtel customer base in Africa exceeds 20 million
JULY 2007	MTC announces intention to close Amsterdam office and to rebrand as "Zain"
SEPTEMBER 2007	Mohamed Ibrahim steps down as chairman to run the Mo Ibrahim Foundation

ANNEX 2

Dramatis personae

Dr Mohamed ("Mo") Ibrahim, *founder and chairman of Celtel and its predecessor company MSI*

Tito Alai Chief, *marketing officer, Celtel*

Dr Saad Al-Barrak, *CEO, MTC Kuwait*

Rick Beveridge, *VP Operations, Celtel*

Sir Richard Branson, *chairman, Virgin Group*

Lord (Simon) Cairns, *(deputy) chairman, Celtel*

Joaquim Chissano, *ex president, Mozambique*

Moez Daya, *chief technical officer, Celtel*

Joe Gatt, *co-founder, Telecel International*

Tsega Gebreyes, *chief business development officer, Celtel*

Sir Chris Gent, *CEO, Vodafone*

Dave Hagedorn, *business development manager, Celtel*

Felda Hardymon, *board member, Celtel, venture capitalist and Harvard professor*

Gretchen Helmer (now Jonell), *lawyer, Celtel*

Sir Julian Horn-Smith, *deputy CEO, Vodafone*

Omari Issa, *chief operating officer, Celtel*

Charlie Jacobs, *partner, Linklaters*

Thomas Jonell, *project director, Celtel*

Alan Knott-Craig, *CEO, Vodacom*

Mamadou Kolade, *business development director, Celtel*

Martin de Koning, *director corporate communications, Celtel*

Kamiel Koot, *CFO, Celtel*

Emily Macauley, *MD, Celtel Burkina Faso*

Strive Masiwya, *founder, Econet Wireless*

Naushad Merali, *chairman, Celtel Kenya and board member, Celtel*

Jean-Marie Messier, *CEO, Vivendi*

Jay Metcalfe, *board member, Celtel*

Sunil Mittal, *founder, Bharti*

Phuthuma Nhleko, *CEO, MTN Group*

Rob Nisbet, *CFO, MTN Group*

Fred Pichon, *company secretary, Celtel*

Marten Pieters, *CEO, Celtel 2003–2006*

Lord (James) Prior, *board member, Celtel*

Cyril Ramaphosa, *chairman, MTN Group*

Terry Rhodes, *co-founder, Celtel*

Sir Alan Rudge, *board member, Celtel and CEO, Celtel 2000–2002*

Miko Rwayitare, *co-founder, Telecel International*

Dr Salim Salim, *board member, Celtel*

Naguib Sawiris, *founder Orascom, Telecom*

Prince Al Waleed, *global investor*

Sir Gerry Whent, *founder, Vodafone*

ANNEX 3
Celtel's financings

EQUITY

Date	Share Price ($)	Amount ($MM)	Total Paid in Capital ($MM)	Lead Investors
Mar-98	2.06	11.1	11.1	Ibrahim, Bessemer, Staff
Jun-98	3.75	5.0	16.1	Ibrahim, Staff, Del Sol
Jun-99	5.5	35.2	51.3	Worldtel, Zephyr, IFC
Feb-00	12	63.0	114.3	Citigroup, EMP, Palio, Alba
Jan-01	15	56.2	170.5	Various, including Staff
Aug-01	15	72.5	243.0	CDC, IFC, Old Mutual, DEG
Aug-03	17	15.0	258.0	FMO
Dec-03	17	62	320.0	Capital Group
Jan-04	17	93.5	413.5	Existing shareholders

Source: Company information.

SHAREHOLDERS

Name	Description
Actis	Formerly part of CDC, the British Government's development finance agency; now an independent fund investing in emerging markets.
African Merchant Bank	South African merchant bank.
AIG Infrastructure Fund	Managed by EMP, the largest private equity infrastructure manager in emerging markets.
Bessemer Venture Partners	The investment arm of one of the oldest funds in the USA.
Blakeney Management	Fund management in emerging markets.
Capital International	Part of the Capital Group, a leading American investment group.
Celtel staff	
Citigroup	The global financial services group.
Communication Venture Partners	Specialist mobile telecoms venture capital firm.
Corporation Financiers Alba	A Spanish holding company forming part of the Grupo March, one of Spain's leading business and financial groups.
DEG	The German development finance agency.

(continued)

SHAREHOLDERS (*continued*)

Name	Description
FMO	The Dutch development finance agency.
Fonditel	A Spanish fund, part of Telefonica.
Ibrahim Family Trusts	Representative of the interests of Celtel's Chairman, Mohamed Ibrahim, founding shareholder of Celtel International.
IFC (International Finance Corporation)	The private sector arm of the World Bank and the largest source of private equity for developing countries.
Old Mutual	The leading fund management company in South Africa.
Palio	A Swiss technology investment company.
Standard Bank of London	The head office of the international banking activities of the Standard Bank Group, Ltd.
Zephyr Management, L.P. Fund	American emerging markets investment fund now part of Prince Al Waleed's Kingdom Holdings.

Source: Celtel International, Annual Report 2003, pg. 33.

DEBT

Date	Prime Lender	Amount ($MM)	Cumulative Amount Raised ($MM)
Jun-98	ING	15	15
Sep-98	CDC	22.5	37.5
Jun-00	CDC	25	62.5
Aug-00	ING et. al	50	112.5
Feb-01	ING	25	137.5
Feb-01	SBL	20	157.5
May-01	CAI	20	177.5
Jun-01	Vivendi	45	222.5
Dec-01	SBL	20	242.5
Aug-01	Moponi	18.4	260.9
Aug-01	FMO	10	270.9
Mar-02	Belgolaise	10	280.9
Jan-03	ING	10	290.9
Feb-03	ING et. al	109	399.9
Feb-03	EAIF	8	407.9
Apr-04	ING (acquisition finance)	70	477.9

Source: Company information.

Index